Contents

INTRODUCTION

Hailstones the size of cricket balls, storms of hurricane proportions and heatwaves that kill thousands. These could be a horrific vision of the future, but they actually happened over a
hundred years ago. Britain has always been a land of fearsome
weather. Perched on the edge of a vast continent, these islands
are a battleground between the fury of the Atlantic on one side
and Europe blowing hot or cold on the other. Sand can shower
down from the Sahara, or snow fall in June on Arctic winds.
Tornadoes have blasted through cities, flashfloods ripped apart
towns and seaports have vanished under the sea. And Britain
holds the world record for the worst episode of air pollution
ever recorded. But the lessons from the past show that rarely
are we ready for the worst of the British weather.

This is more than just a list of awesome extremes, though.
The weather has helped shape our history, fighting off invaders
and stoking up rebellions. It has set off major social upheavals,
helped bring down governments, changed laws, shaped the
economy, and inspired artists and writers. And British weather is
a killer, from heatwaves or cold, winds or floods, that has caused
spectacular accidents at sea and on land, and led to engineering
disasters. And yet vast engineering projects have been needed to
help keep the worst excesses of the weather under control.

We have always been at the mercy of the weather, although things were far worse in olden days, when bad weather led to failed harvests and starvation, or storms flooded huge swaths of land. But even in today's air-conditioned, centrally-heated world, the weather still causes huge misery – just look at the havoc from the summer floods of 2007 or the destruction left behind in the October 1987 storm.

There is also the bizarre side of British climate, where Rickmansworth in the Home Counties is one of the coldest places in Britain, whilst Colwyn Bay in North Wales is a paradise of warmth in winter. And there are fabulous freak phenomena, ranging from balls of glowing light, showers of coal falling from the sky, and even a blue moon.

The British weather has always been a source of fascination, endlessly changing, often on a rollercoaster of extremes. Maybe that's why we have always been a nation of weather anoraks, with a folklore rich in weather forecasts and the first in the world to start our own meteorological office. As Samuel Johnson wrote: 'When two Englishmen meet their talk is of the weather,' which he was probably inspired to write during an infamous summer in June 1758, in which he lamented:

> 'The rainy weather, which has continued the last month, is said to have given great disturbance to the inspectors of barometers... many coats have lost their gloss and many curls have been moistened to flaccidity.'

Small wonder the weather has always been a national obsession.

But never before have we faced such an uncertain time with our weather and climate. Gone are the days of carnivals on the frozen Thames. Now weather records are tumbling with inde-cent haste as the climate turns warmer. Scotland's ski industry is running out of snow, gardeners are mowing their lawns through the winter, and no longer is English wine a joke as French cham-pagne houses buy up vineyards in southern England.

Many question whether this is just a blip in the country's long history of wild weather, or whether it is truly a profound lurch in the climate never witnessed before. Does a searing hot summer mean that we are in throws of global warming, or is it a normal part of our variable climate?

To answer that we need to understand the past, which is why weather records have never been so important. And thanks to our national obsession with the weather, Britain has the longest weather records in the world. Temperature read-ings for central England stretch back to 1659 and rainfall fig-ures for England and Wales go back to 1766. And now weather historians are delving deeper, through diaries, accounts, chronicles and ships' logs for more clues to the past weather.

Each piece of historical evidence means very little on its own, but like bits of jigsaw puzzle they fit together into an un-mistakable picture, that something profound is changing in the character of our present climate that is diverging from the past. Yes, the weather will always blow hot and cold, wet and dry, and that is how the climate has behaved throughout history thanks to natural events such as the effects of volcanic eruptions. But now something different is happening, which is no longer natu-ral, as the long-term trend is growing increasingly under the influence of man-made effects. That is climate change.

So be warned if you thought British weather was just sun-shine and showers. Welcome to *Since Records Began*.

SPRING

Hailstorm Record

1697

Hailstorms of incredible savagery blighted May 1697. Edmund Halley, best known for his comet, recorded a barrage of huge hailstones on 10 May whilst he was in charge of the National Mint at Chester Castle. The storm smashed roofs and windows in a trail of destruction stretching some 53 miles (85km) across North Wales and northwest England.

On 15 May another hailstorm struck of even greater ferocity, in what is believed to be the most violent hailstorm in recorded British history. Although the track of this storm was shorter, from Hitchin to Great Offley, Hertfordshire, it hit much harder. Monster-sized hailstones, described as big as a man's hand, reaching some 5.6in (140mm) across, came down in barrages like cannon balls. Crops were pulverised, the ground looked as if it had been bombed and great oak trees were split apart. Many people were badly injured and one person was reported killed, although it is difficult to believe that many more were not fatally struck down under such an onslaught.

The weather over the rest of the year was also punishing. Another violent hailstorm struck Herefordshire on 17 June, and the summer was thoroughly sodden. Severe frosts and snowfalls began early in November and heralded a savage winter. In fact, the 1690s were blighted by bad weather, possibly caused by a veil of dust blocking out sunlight following huge volcanic eruptions in Iceland and Indonesia.

Armies in the Sky

Aurora 1716

England in 1715 was in political turmoil, as tensions between Jacobites and Hanoverians boiled over into armed conflict. The Jacobites were intent on ousting the Hanoverian king, George I, and putting James Edward Stuart on the throne, to be crowned James III. In the autumn of 1715 a Jacobite rebellion rose up in the North, and the Earl of Derwentwater based in Northumberland led an army into Lancashire to confront Hanoverian forces. But the Earl was soundly beaten at Preston, and was captured and tried for high treason. On 24 February 1716 he was executed at Tower Hill, London, and several days later the hearse with his remains was taken back to his home in Northumberland. But on the evening of 6 March, a quite astonishing spectacle appeared in the sky, as described by an old family servant of Derwentwater:

'A most Beautiful glory appeard over ye hearse, wich all saw, sending forth resplendant streams of all sorts of colours to ye east & west, the finest ever I saw in my Life. It hung like a delicate rich curtain & continued a quarter & half of an hour over ye hearse. There was a great light seen at night in several places & people flockt all night from durham to see the corpse.'

All over the country people were amazed by stunning displays of lights slowly dancing across the night sky, in visions that soon became known as 'Lord Derwentwater's Lights.' In London thousands came out on the streets, and some said that they saw two headless men fighting with flaming swords in the night sky. Another eyewitness described a battle of armies in the sky over Lincoln's Inn Fields.

'All the People were drawn out in the Streets, which were so full one could hardly pass, and all frightened to death... Then appeared in the sky two great Armies which contained thousands of Men and Horses, these seemed fiercely to Encounter Each Other, and the Battle seemed long and doubtful, as if Fortune was in debate with herself... For it was no Meteor or Vapour raised from a Natural Cause, but Something Supernatural, Strange, frightful, Astonishing and Amazing, as is testified by Thousands of People.'

The battle in the sky was no isolated flight of fancy because several people in Oxford saw ' Swords drawn, and Armies fighting in the Air' and in Upton-upon-Severn, Worcestershire, others spoke of armies with weapons in full use, and the smell of the gunpowder and sulphur filled the atmosphere.

But this battle in the sky was an aurora, best described by another eyewitness in London:

'The display appeared at first like a huge body of light, compact within itself, but without motion; but in a little time it began to move and separate, extruding towards the west, when it seemed to dispose itself into columns or pillars of flame... and after many undulatory motions and vibrations, there appeared to be a continual fulguration, interspersed with green, red, blue, and yellow.'

Natural phenomena in the sky were often seen as symbols of conflicts closer to home, and in such turbulent times as 1716, the armies fighting in the sky carried great significance.

Although auroras are rare in Britain, and especially southern England, they are seen occasionally. Over 200 years later, another spectacular display was seen. In January 1938, the west Midlands was treated to a two-hour aurora, and it is interesting how people in the 20th century described the sight. One eyewitness reported in the *Wolverhampton Express and Star* that the sky was filled with 'bright red, luminous, feathery clouds.' Another likened the deep, red glow in the sky to the reflection of the glare from a blast furnace fire. Another thought there was a big fire at a local colliery and phoned the fire brigade. However, there were some who saw it in apocalyptic terms: 'In some quarters it was said the world was coming to an end,' the paper added. The huge geomagnetic storm in the upper atmosphere that created the aurora was also responsible for delaying express trains on the Manchester to Sheffield line after power disturbances hit the signalling apparatus. The aurora returned later in the year, on 12 May 1938, appearing as a brilliant purple colour.

More recently, an outstanding aurora display was seen on 29 October 2003. Reports of fantastic coloured lights came from north Scotland to Land's End, although for much of the country thick clouds blotted out the view. A green glow was reported first from Co. Meath, Eire, then swept eastwards and at Colerane, Northern Ireland, Mike Tullett, a retired meteorology lecturer saw the entire northern half of the sky light up. 'It was so spectacular, as if streaks of light came down to the ground with the colours of the rainbow, rather like shimmering curtains,' he remarked. 'It was so bright that at times the stars were obliterated.'

Normally the Earth's magnetic field shields us from storms erupting out of the Sun, but intense solar storms full of electrical particles can tear our magnetic defences apart and penetrate deeper into the upper atmosphere. There the electrical charges set the gases glowing in different colours, rather like a neon sign lighting up. The polar regions usually get the best auroras because the Earth's magnetic field is like a huge bar magnet, drawing down the solar particles into the magnetic poles. Only the most intense geomagnetic storms give auroras over the whole UK.

These same violent upheavals in the Earth's magnetic field also affect satellites and communications, and two days after the 2003 aurora a Japanese communications satellite was badly damaged by the solar storm. High frequency radio communications used by aircraft over mid-ocean suffered interference problems, and power grids in northern latitudes had to transfer their power loads as the geomagnetic storms induced currents in high voltage cables that could blow up electrical equipment.

Sheffield Dam Burst

1864

In the mid-1800s, Sheffield was a major centre for steel production in the thick of the Industrial Revolution. Huge volumes of water were needed for the booming industry, and Dale Dyke Dam was built a mile long and a quarter mile wide, high on the city's outskirts. By February 1864 the dam was completed and a month later it had filled to its maximum water volume.

On 11 March, a storm raged across Sheffield, whipping spray off the reservoir in sheets of water. A workman returning home that evening was shocked to see a large crack on the 100ft high earthen banks of the reservoir. He ran down the valley for the chief engineer, John Gunson, who inspected the dam and thought the crack was merely a surface problem caused by frost damage or settlement of the earth. But he had the presence of mind to order the water level in the reservoir to be lowered immediately using gunpowder to blow open a hole in the side of the dam and drain off water quickly. It took several attempts to detonate the explosive in the wind and rain, but eventually it worked and the drainage began. However, shortly before midnight a new problem appeared, as Gunson felt a violent shaking of the ground. He looked up at the top of the dam and saw 'water running over like a white sheet in the darkness.' Seconds later a large section of the dam's mighty banks collapsed and a mountain of water was unleashed onto the Loxley valley below. 'The water demon leaped with a voice of thunder from his oozy bed, and rushed with headlong fury down the gorge below,' described *The Illustrated London News*. Some 700 million gallons tore down into Sheffield in a wall of water over 50ft high. According to Robert Rawlinson, the government inspector at the inquest that followed the disaster: 'The force of the water was tremendous and almost inconceivable. The velocity of the wave was awful, and not even a Derby horse could have carried the warning in time to have saved the people down the valley.' Stone houses were torn from their foundations and swept away, generations of families lost, most people died asleep in their beds.

After about 30 minutes the flood gradually subsided leaving a trail of destruction more than 8 miles (13km) long, described as 'looking like a battlefield'. Over 4,000 houses, 106 factories, mills and workshops, 64 other buildings and 20 bridges were completely or partly destroyed. It is estimated that around 240 people were killed in the flood in the biggest dam disaster, and one of the worst man-made disasters of any sort, in British history.

The disaster inquiry concluded that the construction was to blame, and that a small leak in the dam's earthen wall had grown rapidly until the dam failed completely. There were around 7,000 claims for damages against the Sheffield Water Company, and formed one of the largest insurance claims of the entire Victorian period. Following a special Act of Parliament, compensation of £273,988 was paid for damage to property, injury to persons, and loss of life – one of the largest insurance awards of its time.

The Sheffield dam burst was not an isolated accident. A remarkably similar disaster had struck before, in February 1852, after days of heavy rains swamped the large Bilberry reservoir, which supplied mills along the deep and narrow Holme Valley,

Holmfirth, West Yorkshire, best known these days for the setting of the TV series 'The Last of the Summer Wine'.

On February 5, rain fell in torrents all day and by night-time water was seen spilling over the top of the dam. Cracks opened up in the dam's earth embankments before they gave way. 'A loud thundering crash and the pent-up waters rushed with fearful velocity,' reported *The Halifax Guardian*. 'The waters raged down the valley, destroying mills, houses and cottages.'

The floodwaters swept up rocks and boulders, some reported to be at least 5 tons, and bulldozed through Holmfirth lying miles below. Buildings, bridges and even a graveyard were all torn up and swept away. Next day the hills were scoured so deeply by the floodwaters they 'looked like the bed of a mighty river, with just chimneys left from some of the mills.' Holmfirth and the valley for many miles beyond was a scene of devastation that left 81 people dead.

Government inspectors found that the reservoir was poorly designed and built, but no one was prosecuted. Despite this, and several other dam disasters, it was 78 years before legislation was brought in to improve dam safety.

As *The Illustrated London News* commented after the Sheffield flood: 'Another catastrophe of this kind will be a disgrace to the public authorities. It is not, perhaps, a fit matter for legislation; but periodical Government inspection is loudly called for. ... the same mistake may have been made in regard to all the larger reservoirs of water-work companies in the kingdom. 'Hence why the Sheffield dam burst became known as the Forgotten Flood.

But in 1872 November, a dam burst in Halifax, killing ten people. And on 2 November 1925, a small dam in North Wales burst after weeks of heavy rain and pulverised the village of Dolgarrog, killing 16 people. Even the village church was swept away, and survivors said they could hear the bell ringing as the church disappeared. It was later found that the manager and directors of the company that owned the dam knew there were serious defects in it but kept it secret. However, the disaster at Dolgarrog was the turning point, and led Parliament in 1930 to pass new safety laws for reservoirs. These laws were updated in 1975 to cover the safety of all reservoirs in the UK containing over 25,000 cubic metres of water, which have to be inspected every ten years, or more often, and makes reservoir owners legally responsible for their safety.

Since the Dolgarrog disaster no more lives have been lost in the UK from a dam burst, but there are fears that disasters could still occur. Smaller dams and reservoirs under 25,000 cubic metres fall outside the safety legislation. Most of these are over 100 years old, built of earth, and many are near built-up areas, so there is potential for another disaster. A small reservoir in a quarry near Derby failed, and in June 2005 another small dam failed at Boltby, North Yorkshire, although both caused no significant damage. 'The problem is that legislation is concerned with the volume of reservoirs, rather than the risk they pose,' explains Andrew Hughes of the British Dam Society. 'They're often old, left over from the Industrial Revolution in the Pennines. In my opinion, these dams could put lives at risk, as well as damage buildings.'

March Blizzard

1891

March is supposed to bring the first days of spring, but sometimes winter refuses to leave. In March 1891, southern England and Wales was plunged into a storm of such cold and ferocity that it became known as The Great Blizzard, of which great tales were told.

The storm raged for four days, and the winds were so violent that over half a million trees blew down, roofs were ripped off houses and dozens of ships were wrecked in the English Channel. The West Country was worst hit, where snowdrifts piled up to 20ft (6m) high, burying houses, roads and railways and even settled in church spires and muffled their bells. One gigantic snowdrift was estimated at around 300ft (30m) deep and filled a ravine at Tavy Cleave on Dartmoor, over which a vicar was reported to have skied from one side to the other.

'No such storm had visited the West of England within remembrance,' *The Times* exclaimed. Several people froze to death outdoors and thousands of livestock perished. But the worst casualties were at sea, as *The Times* reported on 14 March: 'Coastlines were strewn with the wreckage of boats and ships along the Southwest coast, Dover, Folkestone, Hastings and Dungeness... boats had sails torn to shreds.' Some 220 lives were lost in 65 ships wrecked in the English Channel, one of the worst maritime disasters of the century.

Perhaps the greatest tragedy was the *Bay of Panama*, reckoned to be the finest sailing ship of its time: sleek, fast, with a modern steel hull and four square-rigged masts. She was bound for Dundee with a cargo of jute from Calcutta and had made swift progress before sailing into the English Channel off Cornwall. But as the weather deteriorated and visibility closed in, the Captain, David Wright, grew uncertain of his exact position and took depth soundings close to shore. But during the night the blizzard overwhelmed the ship and drove her onto the rocky shoreline under cliffs near Porthallow, Cornwall, only 65ft (20m) from the shore. The crew set off distress flares but they were completely blotted out in the blinding snow. As the *Bay of Panama* became jammed against rocks she listed badly before waves ripped her open, drowning Captain Wright and many of his crew. Those that managed to survive the onslaught clung for their lives onto the rigging, but several men froze to death encased in ice, or else drowned. Next day, a local farmer discovered the ship and a rescue was launched using a rocket line shot aboard the ship, and 17 men were rescued out of the original crew of 40.

Inland, the snow fell so thick and fast that many trains were buried for days in monstrous drifts. *The Times* reported the harrowing experience of one train driver snowed in on Dartmoor. 'We ran into a pile of snow 7ft high and about 100 yards long and we were fairly caught,' described engine driver John Murray. 'A terrific snow was then raging and we could see only a few yards ahead. In about one hour I became insensible.' He and his mate were found suffering frostbite and taken to shelter. 'Had we not been rescued just in the nick of time there is no doubt we should have been frozen to death,' he said.

One of the most astonishing railway ordeals was the 'Zulu' express train that left London for Plymouth at 3pm on 9 March. Later that evening it was blocked by a huge wall of snow on

Dartmoor, and conditions on board the train steadily deteriorated as snow began to blast its way into the carriages and the heating ran out. It took two days before a farmer in a house only about 200 yards away spotted the train sticking out of the snow, and even then he had to dig his way through to rescue the stranded passengers who were led to a local village and put up in houses. Altogether, it took 300 navvies to rescue the train, which eventually arrived in Plymouth eight days late, something of a record even by today's standards. 'The guard, drivers and two or three passengers had kept with the train the whole journey,' reported *The Times*.

Other trains were lost in the snow and railways blocked for up to a fortnight. Lady passengers on the Princetown to Plymouth train were reported to be very distressed when the train became buried, carriages filled up with snow right up to the hat racks despite the doors being shut tight, and the engine driver announced 'We ought never to have started'. The passengers spent 36 hours trapped in the train before finding refuge on a Dartmoor farm.

It was indeed a wretched spring, and things seemed to get from bad to worse. The following Whitsun, 18 May, the entire country felt the full wrath of another blast of winter with heavy falls of snow, hail, rain and frost. Orchards were devastated, lambs killed, ponds froze over and there were reports of a flu epidemic. In fact, it was so cold that some snowdrifts across Dartmoor remained visible until June.

The Long Drought

Spring 1893

'The spring is more forward than has been known for many years. It is to be feared, however, that the excessive drought is doing serious injury to grass lands and green crops of all kinds; and in many places there is a great scarcity of water.' This report sounds very familiar, but it actually was written in April 1893 in Appleby, Leicestershire during the longest drought on record in Britain.

There was no rain recorded whatsoever in many places from the end of February to mid-May. Mile End in East London broke the UK record for the longest continuous run without rain: 73 days, from 4 March to 15 May. Perhaps most remarkable of all, the drought stretched across most of Britain, except for western Scotland.

It was also the warmest spring on record. Spring flowers bloomed unusually early – a weather observer in Addington, Kent was amazed to see his plum trees in full blossom in the last week of March. Hot sunshine blazed down and for seaside resorts the holiday season began ridiculously early, attracting thousands of day-trippers. Eastbourne basked in some seven hours' sunshine a day in March alone, and Torquay recorded sunshine every day from April to June, notching up 93 per cent of the maximum sunlight possible. In fact, 1893 was one of the sunniest years on record across southern England.

However, the high pressure systems that gave the clear blue skies during the day also gave some very chilly nights, with frequent frosts and fogs. And, of course, the high pressure also kept rain away that helped stoke up a drought.

It was one of the driest Aprils ever known, with many farmers unable to sow cereal crops. By May, plants were becoming scorched, the ground was parched, meadows burnt dry, root crops collapsed and the hay crop was a disastrous failure. 'This truly remarkable season makes one think summer is over; gooseberries are ripe, currants ripe and full coloured. What flowers shall we have for the summer when it comes?,' described a weather observer in Haverfordwest, Wales. Large-scale wildfires broke out on heaths and woodlands, and plagues of wasps appeared in mid-May.

One correspondent in Norfolk wrote in the *Daily Telegraph*:

> 'In many villages, quite a water famine is being experienced, and upon the meadow lands which were closed for hay, the crop has been lost and it has become necessary to pasture thereon cattle, sheep, lambs. Only exceedingly low prices for stock can be obtained, owing to the scarcity of fodder. In many districts a blight has befallen tree and low fruits. In some places corn looks well, in other parts extensive tracts are withered. Barley and oats can only be half a crop at best. Roots and other crops look like a failure.'

Life in the cities became increasingly unpleasant. 'A curious indication of the long drought has been observed in the neighbourhood of Fleet-Street, swarms of rats having invaded some of the lanes leading down to the Thames,' *The Times* reported on 15 May. 'The sewers are apparently dry, notwithstanding the liberal flushings which the streets receive in the early hours of the morning, and the rats are famished. They have been described by an old inhabitant of Fleet-Street, who caught 17 in one night, as like rabbits in a warren. This is a new form of plague, although not so troublesome as the mosquitoes which made life miserable in Fleet-Street last year.'

Part of the problem was that by the 1890s a new system of fresh water was piped to some half a million households in London, and with WCs to flush and baths to fill, most people took it as their right to have as much water as they wanted. But with the drought the supplies of water dwindled and the private water companies that supplied the water asked their customers to cut down on water use, a familiar cry over a century later. It did not seem to make much difference, though, and water shortages steadily worsened as reservoirs dried up. The water companies had monopolies in each of their own areas, so it was easy for them to ration water in what were called 'water famines' that lasted for weeks in the East End. Indignant customers accused the water companies of causing the shortages from burst water pipes, and using the drought for raising prices and profiteering. The East London Waterworks Company on their part stressed the 'unusual drought', which happily exempted them from penalties for failure to supply water, and pointed to consumer 'waste'. The company cut water supplies for several hours each day, saving an estimated 10 million gallons a day, it was claimed. Even though the company was supposed to publicise the hours of supply, many customers were left uninformed, so to cope with the erratic supplies, many people simply left their tap open with a bath tub underneath, ready to collect water whenever the water came back on. *The Times* suggested 'those 10 million gallons tell an eloquent tale of kitchen taps and garden hoses left running all night'.

But many people were forced to drink water from the cisterns of their WCs, and those without their own WC used pots, jugs, buckets or any other container to store water. Small businesses that depended on water, such as laundries, were also hit badly by the shortages.

Although the spring of 1893 was record-breaking, it was not that unusual and was only part of a run of dry years stretching from the mid-1880s to 1909, a remarkable period called the 'Long Drought'. Some of those years actually had wet summers, but they still suffered water shortages through lack of winter rains. It is the rains from November to April that are so critical, because this is when rains replenish most of the underground reserves, rivers and reservoirs, without losing water by evaporation from the ground, trees and plants.

The effects of the Long Drought were stark. In 1884–85, water supplies were so restricted in many towns in England that water was sold by the bucket in some places. In one emergency, water had to be delivered by train in milk cans to parts of Lancashire in August 1887 and served out to people by station masters. That same summer thousands of men were thrown out of work in quarries and tin works in North Wales when local reservoirs ran out of water. As rivers dwindled they rapidly turned into sewers, and at Mountain Ash in Wales typhoid broke out. The lack of water in rivers also brought Lancashire mills and workshops to a standstill, barges could no longer work along canals and boats ran aground in rivers.

During the 1890s there was a run of four successive dry winters, the lowest rainfall in a 123-year drought record, and especially bad in the English lowlands. There has been no drier September–April period recorded since for England and Wales.

In 1895, a ten-week drought left wells without water and reservoirs sank to staggeringly low levels. That brought even worse water restrictions, with more than half the houses in West Ham, London without water. 'The drought is very serious. The ponds on the high grounds are dried up, and the cattle have to be watered by carts. Hay will not be half the average. And cherries, plums, damsons, and apples have suffered', reported *The Daily Mail* of Slough on 6 June 1895.

'A remarkable drought is being experienced in Devon and Cornwall. And the water supply of Plymouth and other towns has become so short that even water required for domestic purposes is only turned on for a limited period. With the exception of a slight rainfall at the beginning of last month the drought has lasted since the middle of March. Agriculturalists are suffering severely,' reported the *Bristol Times and Mirror*, 1 July 1895.

The insanitary conditions caused an alarming increase in infant deaths, mostly from diarrhoea. There were also fears of a lack of water for fighting fires when many street fire hydrants, which by law had to be supplied with water, were found to have run dry, and firemen had to wait for supplies to be turned on.

The Long Drought was a devastating blow to agriculture in what was already a great agricultural depression that had begun in the 1870s with a run of wet summers that ruined harvests. As British grain became scarce and costly it became cheaper to import American grain, many farmers gave up growing cereals and switched to livestock instead. But by the 1880s imports of refrigerated and canned meat and dairy food hit British livestock farmers hard. So, the long drought of the 1890s tipped farming into an even steeper decline –

agriculture's contribution to the national output fell from one-sixth in 1867–69 to less than a fifteenth by 1911–13. Farmers simply gave up their farms, and farm rents collapsed in the arable areas of eastern and southern England.

However, there was another reason for the water shortages and, again, it was a remarkably familiar problem that resonates in modern times. Huge amounts of water were lost through leaking pipes. Many of the metal water mains in streets were fractured by deep penetrating frosts in the bitter winters of the 1890s. In May 1895, Reading alone had 915 fractured main water pipes, and in Sheffield 170,000 houses were left without water from burst pipes.

The following year, 1896, from January to June there was less than half the normal rainfall. During August 1896 households in East London were restricted to four hours' water a day but the East London Company insisted that the problem was 'consumers took not the slightest interest in the... careful use of the water and made no provision against drought, frost, or the breaking of the mains', reminding those who left their taps running that 'this waste is distinctly illegal and... a great source of inconvenience to neighbouring consumers'. In fact, the company encouraged their customers to inform on neighbours who wasted water.

Research by Frank Trentmann and Vanessa Taylor at Birkbeck College, London University reveals that the droughts and deprivations of the 1890s led to new consumer movements, not that dissimilar to modern-day consumer campaigns. In the East End, one draper took his water company to court for failing to give the statutory 'proper supply of water for domestic purposes'. He lost but his action mobilised others to stand up for their rights. Whereas in previous decades the lack of basic life necessities often sparked off riots, the water consumer movement turned to political activism for a remedy. The East London Water Consumer Defence Association was formed and demanded water companies be taken over by local municipals. They held public meetings, lobbied the press and politicians, set up advice bureaus and provided legal support, and their pressure led to reduced rates and settling of disputes.

In truth, the water companies were struggling to cope with rising demands, with or without droughts. The constant supply of piped water brought a huge upsurge in demand for WCs and baths, plus the populations of the major cities were booming in any case. Eventually in 1902 the London water companies were taken over by municipals, but it did not solve the underlying problems. Shortages of water continued long afterwards, in severe droughts in 1921, 1929, 1933–34 and the 1950s. As the rising population grew increasingly affluent, more water was demanded by consumers for WCs, baths, washing machines, gardens as well as swimming pools and golf courses, whilst farming increasingly demanded more water for irrigation. But the same accusations were thrown around, as *The Times* commented on 6 December 1952: 'The average consumer wastes water with unthinking abandon; it has become usual to regard heavy water consumption as a virtuous sign of civilisation and super-hygiene.'

It seems little has changed since the Long Drought. We are still vulnerable to long droughts, as shown in the two unusually dry winters in 2004 and 2005. Southern Britain struggled to cope when these dry winters were followed up by a blistering hot summer in 2006. In the Thames area it produced the greatest water deficit since the early 1890s and led to water restrictions affecting 13 million consumers across southern England. Low river flows, limited sewage disposal and the drying-up of streams and ponds brought on serious pollution problems that also killed fish.

Since the 1890s, water demand has soared more than five-fold, water pipe leakage continues to be a huge problem, water companies still complain about wastage by consumers, and domestic customers and farmers demand even more water. All of which puts a huge burden on water supplies. But now a new problem is emerging, as summers grow hotter and more arid as climate change takes hold. Small wonder that in 2007 the Mayor of London, Ken Livingstone, proposed a water desalination plant on the Thames Estuary to supply Londoners with much needed fresh water.

However, a recent study of the history of droughts in Britain suggests that over the past 50 years we have enjoyed a largely benign period of climate. In fact, prolonged dry periods are not unusual for our climate, and if we have a repeat of the Long Drought it could set off a crisis in modern-day Britain. As Terry Marsh, hydrologist at the Centre for Ecology and Hydrology, explained: 'We could see drought patterns return to the sort of variability we've seen before. The surprise element is that most of us have grown up in a situation where winters have been relatively mild and wet, certainly in the past 30 years.' The warning is loud and clear, but what can be done about drought remains to be seen.

Smog, London and Monet

1904

On 27 May 1904, London descended into an intensely dark gloom, with gas lights and candles lit at midday. People reported unusual oscillations in their barometer readings, but what exactly caused this strange darkness is not certain. Probably it was smog from coal smoke. London was the most polluted city in the world, known as 'The Big Smoke' as a million coal fires and countless factories, railway engines, barges and boats gave off smoke and sulphur dioxide. When the sulphur and smoke particles became trapped near the ground it conjured up a classic London peasouper, made famous in Sherlock Holmes novels. But the horrors of the Victorian smogs that enveloped the cities of Britain ruined public health through bronchitis, pneumonia and emphysema, and rickets brought on by lack of sunshine.

Smogs were usually worst in the winter, so the gloom of May 1904 was unusual. A claustrophobic gloom gripped the city, but not everyone saw it so bleakly. The impressionist artist Monet was staying at the Savoy Hotel in London where he was fascinated by the weird and wonderful light and colours of the smog. 'Without the fog, London would not be a beautiful city,' he wrote. Monet was particularly attracted to the nearby Houses of Parliament shrouded in smog, and his picture 'Le Parlement, Effet de Brouillard' shows the spires of the building standing up like silhouetted fingers shrouded by an air of smoky purples, blues and greys. It was the atmosphere which inspired him: 'I want to paint the air which surrounds the bridge, the house, the boat, the beauty of the air in which these objects are located,' he explained.

But was this artistic licence, or did Monet catch a real phenomenon in his paintings? In 'Houses of Parliament, London, Sun Breaking Through the Fog', the sun is reduced to a feeble ball of light struggling to break through a purple shroud of smoky fog. Meteorologists Jacob Baker and John Thornes at Birmingham University examined the picture and used the spires and towers of Parliament as markers to calculate the precise angle of the Sun. From this they revealed that Monet made the picture during late afternoon and early evening in February and March 1900, which matched the letters Monet wrote to his wife. 'The position of the Sun images is spot on,' remarked Dr Thornes. The scientists also worked out exactly where Monet stood with his easel, in the grounds of St Thomas' Hospital opposite Parliament.

Another feature of Monet's London pictures was the appalling visibility and this, too, probably was accurate. During the winter of 1901–02, the London Fog Inquiry made daily observations from Victoria Tower on Parliament to find out how bad the visibility was. Over that entire winter the observers never saw St Paul's Cathedral, about 1.5 miles (2.5km) away, and more usually could see only for about a mile or less, which matched the visibility in Monet's pictures. 'This shows these paintings are potentially an accurate visual record of the urban atmosphere of Victorian London,' explained Dr Thornes.

Easter Weather

More and more people are using Easter to flee the UK for warmer climates, and a look back at the past weather records shows this often a good choice. Snow is more likely to fall at Easter than at Christmas, and over the last 50 years snow has fallen on more than a dozen Easters, most recently in 2008. The only saving grace is that snow in April tends to be wet, slushy and often melts away as quickly as it appears.

Possibly the most diabolical Easter was in 1908. Because the date of Easter can fall between 22 March and 25 April, the coldest Easters tend to strike in late March, but not always. In 1908, Easter Sunday fell late, on 19 April, so most people expected decent weather. But a snowstorm and hailstorms struck southern Britain, while Scotland and northern England experienced intense frosts. The vicar of Hawkedon, near Bury St Edmunds, reported 3 inches of snow on Sunday night and another 4 inches the following morning. But the dire conditions did not deter the great cricketer, Dr W.G. Grace, playing his final game at Whistable, Kent on Easter Monday, despite snow on the pitch. He took the field for the Gentlemen of England against Surrey and scored 14 and 25 runs.

But that Easter storm was only a taste of something worse to come. On 25 April a full-blown blizzard swept England with an even greater fall of snow, piled up to 2ft (60cm) high in the worst-hit areas of Hampshire and Berkshire. Oxford recorded 18in (46cm) of snow, the heaviest snowfall in the city's history during the 20th century, just as the new summer term began at the university. Even the Scilly Isles and Channel Islands were snowed under. Greenhouses collapsed under the weight of snow, telegraph poles keeled over, and horse-drawn sleighs were used to haul supplies to villages marooned by snow.

A blinding snowstorm raged in the Solent when the passenger liner *St Paul* left Southampton bound for New York. In appalling visibility, she collided with the cruiser HMS *Gladiator* side-on, tearing a hole in the warship, which immediately took on water and listed. One of the stokers on the *Gladiator* was reported by *The Times* as saying: 'We were thrown clean across the other side of the ship. We could see immediately that we had been run into, for another vessel's bow had crashed through us and one poor fellow was killed on the spot.' The *Gladiator* managed to beach on rocks about 400 yards from a beach on the Isle of Wight, and rescuers on shore waded through the icy waters to reach the crew, but 30 men were lost, most of them drowned.

But one surprising and useful thing did come out of the atrocious weather. It was the day of the FA Cup Final between Wolverhampton Wanderers and Newcastle United, played at Crystal Palace. The omens looked bad when before the match the ground was lashed with rain, sleet and heavy snow showers. But just before kick-off the skies cleared and a crowd of 75,000 watched Wolves win 3–1 in bright sunshine. Something good came of the snow showers, however. One Wolves supporter, Captain Gladstone Adams, drove to the match in his motor-car, something of a novelty in those days. But as he drove through the snowstorm he struggled to see where he was going as snow stuck to his windscreen, and he was forced to fold down his windscreen. That gave him the idea for inventing the windscreen wiper, which he patented in 1911.

Snow is just one of the weather surprises that Easter has up its sleeve. In 1998, storms on Good Friday brought torrential rains – what the Environment Agency described as 'monsoon levels' – in a swathe from Herefordshire to Norfolk, and almost 5 inches of rain fell just north of Banbury. Rivers soon broke their banks and devastating floods hit Leamington, Stratford-upon-Avon, Northampton, Peterborough and Wisbech. To cap it all, blizzards also swept Scotland and Devon and there were reports of a tornado in the northeast of England. The floods forced thousands of people to leave their homes and six people died. The insurance bill came to around £1 billion and a government inquiry was called to discover why flood warnings were so inadequate.

Easter's reputation for wintry weather even had a starring role during a debate in Parliament in 1928 about fixing the date of Easter. It was decided to make the date between 9 and 15 of April, when there was a better chance of good weather. However, at that time Scottish meteorologist Alexander Buchan revealed that over the previous 50 years this period tended to be cold. And to prove the point, that April turned cold again. Even though an Act to fix the date of Easter was passed, it did not come into law and remains in legal limbo until the World Council of Churches agrees to it.

So why is Easter so prone to atrocious weather? Much depends on the jet stream, the high-speed wind blowing above Britain from the west at a height of several miles. In springtime, the jet's path can snake around as the atmosphere warms up, and that can profoundly alter weather patterns below. When the jet wobbles southwards it can drag down cold air from the north, bringing arctic winds and snows. Then, over Africa, it can head north again, pulling in warm Saharan air and a mini-heatwave.

This is also the time of year when the seas around Britain are close to their coldest, after losing their heat over the winter like radiators cooling down. The saving grace, though, is that the lengthening days and strengthening sunlight in spring means that land is warming up, and snow tends to melt away quickly.

A History of Easter Phenomena

1892. Easter Sunday a great snowstorm swept through Kent and in Sevenoaks a boy was prosecuted for throwing a snowball at a local councillor.

1904. In Essex a ball of lightning exploded.

1907. A fireball, probably intense ball lightning, set fire to a girls' school at Bramley near Godalming in Surrey, but luckily the pupils were away on holiday.

1908. A blizzard swept England with deep snow in central districts.

1935. A pair of mock suns and rings of haloes appeared in the sky at Wick in Scotland, an optical illusion caused by high wispy clouds breaking up the sunlight.

1946. A rare and vivid aurora borealis (the northern lights) appeared during the night in Kent.

1947. Heavy sleet and snow fell in Sussex, but soon melted.

1958. Temperatures were bitterly cold and thick snow lay on the ground over Easter.

1963. A waterspout appeared in the sea off Looe in Cornwall.

1964. A tornado struck Warren Street, near Ashford in Kent.

1975. Widespread snow fell. In fact, snow showers fell almost daily from 26 March until 10 April.

1979. Over eastern England the maximum temperatures plummeted 11°C (20°F) from Easter Sunday to Monday. Spouts of water that looked like monsters were seen moving across Earlswood Lakes near Reigate, Surrey, caused by freak whirlwinds.

1981. A deluge of rain fell on Horsham in Sussex. Two weeks later many parts of England had the worst late-April snowstorms of the 20th century.

1983. The most disruptive Easter snowfalls in recent years struck most parts of Britain.

1984. A golfer was struck by lightning just outside Dublin.

1994. Easter fell on 11 April, but the first ten days of April brought gales gusting to 100mph (160km/h), several disrupting snowfalls (1ft/30cm deep in places) and floods in the West Country. Meanwhile in Australia a shower of fish fell on Dunmarra in the Northern Territories.

1998. Torrential rains and devastating floods struck from the Welsh borders to East Anglia.

2000. Torrential rain across large parts of Britain.

2008. Widespread snow.

Derby Thunderstorm

1911

The Derby horse race has been held on Epsom Downs since 1779, usually in fine summery weather. But sometimes conditions have been shocking, and the worst was the Derby of 1911 on 31 May, the Coronation Derby attended by George V and the Queen.

The day had been stifling hot and muggy before a ferocious thunderstorm broke out in the late afternoon. The Royals had already left and other racegoers were leaving the course when cascades of rain and hail crashed down as the sky erupted with a barrage of lightning and ear-splitting thunder. Cars broke down in the floods of water, blocking roads, horses frightened by the lightning became fractious and either refused to proceed or pranced about. The rain poured down mercilessly, turning the ground into mud a foot deep, in which people on foot had to walk. Thousands of people ran to the station on foot, and drenched to the skin; others sought the shelter of the canvas refreshment booths, the floors of which were converted into quagmires.

A marquee on the racecourse was hit by lightning, felling eight people inside; outside, a group of 12 people trying to shelter by a wall were thrown to the ground and two killed when they were struck by lightning. Another lightning strike hurled passengers to the floor or completely out of a horse-drawn carriage, leaving one young man dead. Strangely, one of the other passengers saw what he described as a ball of fire pass through. At a coroner's inquest into the deaths, 'The Coroner said of it he had heard of people say that the storm was judgment upon those who attended the Derby,' reported the *Epsom District Times*.

Thunderstorms raged all across London and the suburbs for much of the evening, causing landslides on railways and flooded streets and stations; 15 people were killed and dozens injured, many of them sheltering under trees. The street opposite St Paul's Station resembled a river and a torrent deluged down Ludgate Hill. Water poured down the Strand and flooded basements and kitchens of several large hotels. Several churches were hit by lightning and Trinity Church, Marylebone was struck during an anniversary service and the congregation evacuated when a portion of the south gallery caught fire. The Royal Exchange in the City was hit and fire broke out. It was one of the worst thunderstorms known in London and the Home Counties.

George V Jubilee

1935

May 1935 saw national celebrations for George V's Silver Jubilee, starting on 6 May with a service at St Paul's Cathedral followed by street parties. The celebrations carried on for most of the month, and although they began in warm sunshine, the weather took a nosedive on 16 May. The notorious frost hollow at Rickmansworth, Hertforshire experienced a bone-chilling night of –8.6°C (16.5°F) with heavy frost, followed in the Yorkshire Dales by heavy snow with snowdrifts 2–3ft (60–90cm) deep, with villagers digging themselves out of their homes. Even normally mild western regions had substantial snowfalls, with Devon and Cornwall said to 'look like Christmas'; in parts of the southwest it was the first snow for over a year. Even the famously mild Scilly Isles suffered heavy snow showers.

'A Return To Winter' declared *The Times*, describing how thousands of spectators braved the biting cold winds to watch the ships of the Home Fleet in the Thames Estuary off Southend for the Jubilee celebrations.

Meanwhile, exhibitors at the Chelsea Flower Show were frantically trying to keep their plants protected in heated greenhouses. Snow stopped play at the Dunlop Southport golf tournament, and the *Daily Telegraph* reported how players had to be recalled from the course. 'Henry Cotton, the open champion, was one of them. He got as far as the second hole, by which time the snow was a quarter of an inch thick. On his way to the first tee he protested that it was ridiculous to play under such conditions. Cotton, in turning to come down the hillside, fell on his back. Fortunately he suffered no ill effects.' Another player, William Laidlaw, took four putts on the first green, his ball gathering so much snow it 'was as big as an orange' by the time it reached the hole. One caddie collapsed from the cold and had to be carried back to the clubhouse by spectators.

Several inches of snow brought racing to a halt at Haydock Park racecourse, and two railway engines were derailed by snow in Yorkshire. Fruit and vegetable growers faced ruin after frost destroyed much of their crops. 'A few days ago the countryside was a sea of blossom; now one night's frost has entailed a loss of thousands of pounds in the Sittingbourne district alone,' reported *The Times*. One fruit grower in Hereford reported: 'There is nothing left at all. The apples and pears are black and the cherries look like bags of water.' But one large apple grower saved his crop by lighting over 2,000 oil lamps over 40 acres (16ha) around midnight to beat off the frost.

Whitsun Flood

1944

The Whitsun holiday, or Spring Bank Holiday as it is now, has had a chequered history of weather. At one extreme, Whitsun in 1891 was so cold that hail, rain and snow fell over much of England and Wales, with 4in (10cm) of snow on the Yorkshire Wolds. 'The potato crop was almost totally destroyed, the fruit blossom blighted and harm to grouse and other young game,' reported *The Times*. 'The wintry wealth of this Whitsuntide prevented the volume of travel recorded last year, and there were far fewer visitors to Kew Gardens and Madame Taussauds.'

Whitsun in 1944 was record-breaking. For most of the holiday there was a sensational heatwave, and by Monday temperatures over much of the southeast had soared to 32.8°C (92°F), the hottest day of the entire year and also the hottest Whitsun in history. At some locations it was the hottest ever recorded for any day in May.

However, in the afternoon billowing cumulonimbus clouds were furiously feeding on the hot, humid air and by the evening terrific thunderstorms erupted over much of Britain. Hailstones as big as eggs fell over Shropshire and staggering downpours of rain flooded many areas.

The worst deluge struck Holmfirth, a small textile town nestling in the Yorkshire Pennines. However, the fast flowing river that helped power the textile mills was also the town's downfall. People were returning home after Whitsun processions, relaxing in the park or in the local cinema, when the thunderstorm broke at 6.30pm and triggered a violent cloudburst. The River Holme quickly swelled, bursting its banks and sending a raging torrent 8ft (2.5m) high through the town. Whole streets were torn apart as roads and bridges were ripped up and factories, shops and houses collapsed. 'Some people were returning to their homes after the Whit Monday processions and singing in Victoria Park, others were in the local Valley Theatre when a flood warning flashed onto the screen,' reported one local resident. Dozens of families were left homeless and three people were killed, trapped in their houses.

However, the disaster was called The Forgotten Flood because it struck only a week before before the D-Day invasion of

Normandy in the Second World War, and there was a news blackout on the weather to avoid the Germans learning about weather conditions in Britain.

Longest Running Tornado

Tornadoes in England are usually fairly weak and only last a few minutes over a mile or so, but some are exceptionally violent.

At 2pm on 21 May 1950, ferocious thunderstorms were raging across eastern England, stoked up by hot, humid air from North Africa. Shortly afterwards on the Chiltern hills eyewitnesses spoke of a dense, black cloud gathering on the horizon and quickly developing into the dark column of a tornado. The tornado swept through Wendover, Buckinghamshire; full-grown elms and walnut trees were felled, roofs were lifted off and a column of water sucked up from a canal.

After Wendover, the tornado disappeared before reappearing at nearby Halton Camp, where a heavy roof on a power station was lifted off. From there it zig-zagged through the countryside before bearing down at 4.30pm on Linslade near Leighton Buzzard, on the Buckinghamshire/Bedfordshire border, and wrecked many of the 300 houses and other buildings as it tore through the streets and surrounding fields. One resident, Tony Birch, described the scene as reported in *The Times*: 'Residents were first alarmed by the screaming of the wind. Its force suddenly strengthening, parked motor-cars were lifted bodily, trees were uprooted, drawn high into the air and dropped a considerable distance away. Houses were stripped of the air slates, windows were shattered, and furniture and bedding soaked by the downpour which followed.' Mr S. Collins said that storm looked 'simply monstrous. There were slates flying about everywhere, and the sky was black.' Dozens of people were made homeless, and relief workers were brought in to help in the devastation.

The tornado began to weaken, but found two new spurts of life over low-lying places near the Ouse at Bedford, and at 6.30pm over the fens near Ely. The funnel cloud was last seen over the sea off Blakeney, Norfolk.

Altogether the tornado travelled 68 miles (110km), the longest recorded track for a tornado in the UK and left a trail of damage estimated at over £50,000, well over £1 million in today's prices.

Fen Blow

May

Dust bowls are usually thought of as disasters that strike the American Prairies, where winds whip across the flat farmlands and blow up storms of dust. But the rich fenland farms of East Anglia suffer their own dust bowls, known as fen blows. In 1955, the region experienced one of their worst dust storms known, following an unusually dry spring and a heatwave at the end of April that left the ground parched and turned the dark topsoil to dust. On 4 May a gale blew up, gusting to 65mph (105km/h) across Norfolk, whipping up the soil into a vast black cloud of choking dust. For two days the winds raged, turning day into night and slashing visibility to just a few yards in some places, 'like driving in a thick fog,' described one driver.

As the fields were stripped of rich topsoil, the surrounding dykes and ditches were filled with fine dust reaching 6ft (1.8m) deep. Homes were blasted with the filthy black dirt that seeped through gaps around windows and doors. 'I am looking out of my window and I can see a wall of dust beyond... the house is filling with black layers of silt,' reported one farmer in the *Eastern Daily Press*. Thousands of acres of farmland were stripped of seedling sugarbeet, carrot, onion and lettuce crops, and tons of valuable soil disappeared into the air.

Since then, many farmers have been trying to defend their land against the ravages of the fen blows by planting hedgerows and trees as windbreaks, but still the threat of soil erosion haunts them. After a long drought in spring 2002, winds gusted to over 70mph (110km/h) on 2 March and tore down power lines, ripped off roofs, and swept up a dust storm. A fisherman by a lake described the approaching dustcloud: 'The first thing that caught my eye was a huge cloud of what I first thought was smoke... within minutes we were having to shield our eyes from all the particles being blown around in the air; visibility had literally gone down to a few metres.' When the dust finally settled, farmers were faced with thousands of pounds' worth of lost seeds, plants, fertilisers and priceless topsoil.

The fens were ancient marshlands originally and after centuries of drainage produced an incredibly fertile peaty soil –

some of the best farmland in Britain. But as that peat turned increasingly dry it was left dangerously exposed to high winds. In fact, this is the most serious wind-erosion problem in Britain, and the peat may well be all gone in only a few decades' time. Added to that, the peat has shrunk since the marshes were drained and much of the land now lies below sea level. This has created a weird landscape of ships and boats sailing by above the surrounding fields, and has left the fens increasingly vulnerable to serious flooding from rising sea levels and rivers bursting their banks.

Chernobyl

1986

On 26 April, 1986 the Chernobyl power plant in the Ukraine exploded. A nuclear reactor caught fire and vast amounts of radioactive chemicals were shot into the atmosphere in the worst nuclear accident in history. The blast killed 31 people, forced the evacuation of 135,000 more and led to a dramatic rise in the number of cases of cancer, especially in nearby Belarus. And the repercussions still linger in Britain.

The radioactive plume was picked up and spread by winds far and wide over Europe, central Asia, the Pacific and even reached the western United States. Six days after the meltdown, gases still escaping from the devastated reactor were so hot from radiation they cut a swath through low-lying clouds to reveal clear sky above.

Parts of Britain were caught in the fallout by an unfortunate set of circumstances. The radiation cloud swept in from the southeast on 2 May but remained out of harm's way high and dry in the atmosphere. Then localised rainstorms washed down the radioactivity over a swath of Cumbria, Wales, Scotland and Northern Ireland. The rain concentrated the radioactive isotope caesium 137 and contaminated the ground. Most soils would have bound up the radioactive material, but the peat in upland areas kept the caesium in solution. It was absorbed by grass roots and passed into their leaves, which were eaten by grazing sheep. A ban was imposed on the movement and slaughter of sheep in upland areas, followed by embargoes on more than a million sheep.

Over the years the radiation levels have fallen but the grass in some of the hills is still contaminating sheep. The flocks are monitored regularly for radioactivity, and contaminated animals banned from markets. Experts believe the radioactivity will carry on for another 15 years before the caesium dies away.

The Boat Race

'My word – a huge flash of lightning and then that enormous clap of thunder which frightened the life out of me,' exclaimed Harry Carpenter commentating for the BBC just before the start of the 1987 Boat Race, after lightning narrowly missed the Cambridge boat.

That day was already blighted by bitterly cold near-gale-force winds that threatened to sink the crews. Cambridge were late coming to the starting line and at almost the exact time that the Race should have started there was a sudden violent squall of wind, rain, hail and a lightning bolt that actually struck one of the BBC relay masts. The stake boats, used for holding the race boats in place at the start, dragged their anchors and the umpire's launch broke down. But the waters calmed down enough to start the race, eventually.

The Boat Race has faced other battles with the weather in its 150-year history, and has become something of a litmus paper for springtime weather. Strong winds sank Cambridge in 1859 and 1978, and Oxford in 1925 and 1951. To even things up, though, both crews sank in 1912 in a gale. Probably the most diabolical race weather was in 1952, when a blizzard produced a whiteout that left the crews struggling to see where they going and battling against gale-force winds. The crews remained neck-and-neck over the whole race, which Oxford just won by a small margin.

In 1958, the crews had a job seeing where they were going when heavy mist and rain descended like a blanket on the river. Spectators on the river bank were left in the dark, but for the crews the tide was good and the water was calm with very little wind.

For the record, the coldest race was −4.1°C (24.8°F) on 7 March 1942 and the warmest was on 29 March 1965 at 23.6°C (74.5°F).

Yellow Dust Showers

A strange shower of yellow dust covered eastern England on 10 May 2006. People woke up to find their windows and cars covered in the fine yellow powder, and reports were especially thick along the east coast. Anyone who washed their windows or car found exactly the same problem reappeared the following day.

At first, it was thought that a shower of dusty sand had fallen from the Sahara following a desert dust storm and dropped down from a great height over the UK. Or it was reminiscent of the strange yellow showers that American soldiers were covered in during the Vietnam, and which some feared was some sort of chemical weapon, but which turned out to be bee droppings.

The Met Office solved the yellow shower mystery of 2006 using satellites. They revealed a large cloud of dust blown on easterly winds across the North Sea from Denmark. This came when there was a tremendous explosion of birch tree pollen in Denmark, the highest levels of this type of pollen since records had begun there 29 years before, and which caused untold misery for Danish hayfever sufferers allergic to the tree pollen.

Each spring birch trees shed their pollen from catkins into the wind in the hope of pollinating flowers on other birch trees. In May 2006, the weather turned warm and sunny and triggered a huge burst of growth in the catkins, which shed clouds of pollen over several days. The mass outpouring of pollen then wafted over the North Sea and inflicted a spring cleaning problem to eastern England.

Oak And Ash

Lawn mowing has given an amazing insight into the changing climate of Britain. Since 1984, pensioner David Grisenthwaite, in Kirkaldy, Fife has kept a diary of when he mows his lawn and reveals that his grass is growing for one month longer than it did 20 years ago.

Mr Grisenthwaite began his diary when the Woodland Trust asked the public to note when they made their first and last lawn cuts each year. These dates gave a good indication of the length of the growing season, and much depended on the temperature at the time.

Although the project lasted a few years, Mr Grisenthwaite carried on with his diary, and by 2004 showed that the first cut in spring was 13 days earlier than in 1984 and the last cut in autumn 17 days later.

So important is Mr Grisenthwaite's lawn mowing record that the Royal Meteorological Society devoted a scientific research paper to it in their journal *Weather*. What amazed the scientists is how the growing season had lengthened in the north of Britain – such a dramatic change was expected much further south. In fact, many people in southern regions of the country reported mowing their lawns all year round.

Ecologist Tim Sparks, of the Centre of Ecology and Hydrology, explained: 'For this to be happening is quite remarkable. You might have expected, from our experience, for this to be happening in the South-West. But if it is happening in Scotland it shows that grass in gardens further south must be growing for at least a month longer.'

However, there have been some unfortunate consequences of the longer grass. Council grass cutters in Carlisle threatened to go on strike at their increasing workload as the grass is growing faster and for longer in the year. The grass-cutting season used to last from April to September, but now extends from March to November. And the grass is growing so fast that the cutters are having to mow an extra 5 miles (8km) a day in the summer. 'People were complaining that the grass looked a real mess because it was growing so long,' explained Jed Craig, the union official for the region. 'But with this extra workload many of our longer-serving workers simply can't cope.' The council tax payers of Carlisle also felt the effects of the longer grass when the council passed on its increased grass-cutting budget.

Another sign of the changing seasons is the race between oak and ash trees to see which one will open its leaves first in spring.

> 'Oak before ash in for a splash,
> Ash before oak in for a soak.'

The folklore saying is supposed to predict the summer: if the oak tree comes in to leaf in springtime before the ash tree then the coming summer will be dry, but if the ash leafs first it predicts a wet one.

However, in the race to see which tree breaks open its leaf first, the oak has been beating the ash convincingly in recent times. It is now leafing ten days earlier on average that it did 20 years ago, but it has got nothing to do with wet summers. It is all because our springtimes are growing warmer, and as oak trees respond more readily to rising temperatures than the ash, so they are in a better position to take advantage of the changing climate.

The ash, on the other hand, is more sensitive to the hours of daylight in spring before it opens its leaves, so it is now caught in a time-warp, trapped in the cooler springtimes of years ago. In the long-run this means that the ash is going to suffer as the oaks shade them in woodlands and gradually squeeze them out.

This was not always the case. In 1736, a Norfolk landowner, Robert Marsham, began recording the first signs of spring-time, and his ancestors continued the work up until the 1950s. Their records show a much closer race between the oak and ash, and also hinted at the origins of the folklore saying. The ash often beat the oak in wetter springtimes, possibly because it has shallow roots, whilst the oak has deeper roots and does much better in drier conditions.

The race between the oak and ash is a symptom of a much greater shift in the British springtime. Migrating birds are returning earlier, nesting and laying eggs earlier. The great tit has adapted its behaviour by laying its eggs around two weeks earlier than 30 years ago. The change enables the birds to make the most of seasonal food: a bonanza of caterpillars that now also occurs around two weeks earlier due to warmer spring temperatures.

Butterflies are appearing nearly a month earlier on average than in the 1940s, and some are emerging as early as the middle of January. For example, the red admiral butterfly is now seen flying around as early as 19 January compared to an average 28 May from the 1940s to the 1970s. Part of the reason is that the red admiral, a summer migrant from the continent, is now overwintering in Britain because it can enjoy the milder winters without wasting energy to fly to warmer climes. In southern England, the *Bombus terrestris* bumblebee has given up hibernation and also has reached as far north as York.

And the change in plants and trees is also having a big effect. The hayfever season used to last from late March to late September but the grass and tree pollen is now a problem from early March to early October.

Helm Wind

There are many parts of the world that have their own special local winds with their own special names – the Mistral blasts down the Rhone Valley of France and the searing heat of the Santa Ana whips up devastating wildfires in southern California. But there is only one wind in Britain that has its own name – the Helm wind of Cumbria.

This powerful, gusting wind roars down the steep slopes of Cross Fell, the highest point in the Pennines, bowling over walkers that dare to venture too close to the peak. According to legend, the Helm wind brought down a Norman army in the days of William the Conqueror, when it threw the French cavalry off their horses and gave the foot soldiers of the local tribes an unexpected victory.

The wind is at its worst in late springtime, when it can last for days on end. But the strangest thing is that the wind fizzles out into a breeze once it reaches the valley bottom a few miles below.

Most spectacular of all, the wind can be seen for miles around as a big bank of clouds hugs the summit of the hills like a helmet – which is how the Helm wind got its name. And several miles downwind a roll of cloud appears like a long sausage, called the 'helm bar', although between these two clouds the sky remains clear.

The Helm wind and its strange clouds was a baffling phenomenon until 1937, when meteorologist Gordon Manley built a weather station and a wooden shed on Cross Fell. Manley regularly stayed in his shed taking readings, even in the bitter cold of winter when the shed could become encased in ice. Small surprise that he found it to be the coldest place in England.

But the Helm wind seemed to blow whenever Manley was not there. Finally after two frustrating years he finally measured the wind and discovered its secrets. It needs an easterly wind to blow across the ramp-shape of the Pennines, rising up the gentle slope on the eastern side before plunging down the steep western face. And like water flowing over a waterfall, the wind crashed down Cross Fell with tremendous force.

As the Helm wind rushed down, Manley likened it to a fast stream passing over a weir that created a standing wave, with the air bobbing up and down over one spot in the valley. This standing wave produced the helm bar cloud, below which the wind stops blowing. The Newcastle Gliding Club later confirmed Manley's discovery, using the uplift of air in the standing wave to soar to 11,140ft (3,395m) a new British altitude gliding record.

SUMMER

The Dangerous Haar

There are few places in the world that have their own special word for fog. But come late spring and early summer, a claustrophobic sea fog known as haar, can turn the east coast of Scotland and northeast England into a depressing gloom. Sometimes it lies menacing as banks of fog offshore, other times it rolls in on easterly winds and can last for weeks on end. Before the cold seas around Britain have had time to warm up, they clash with warm, moist air wafting across the North Sea. The air cools off and condenses into fog or low cloud.

A haar foiled a Scottish invasion of England in 1174. William the Lion, King of Scotland, wanted to expand his kingdom and he invaded Northumbria. His army camped at Alnwick Castle, near the coast, which is now best known for the setting of Hogwarts School in the Harry Potter films. But whilst camped at the castle, William made the mistake of letting his forces spread out looking for plunder. On the night of 11 July, an English force led by Ranulf de Glanvill set out from Newcastle and arrived at Alnwick thick with haar. 'It is said that so dense a fog covered them that they hardly knew whither they went,' described historian William of Newburgh. Under cover of the haar, the small English force launched a surprise attack at dawn on the Scottish encampment at the castle and captured King William. The rest of the Scots fled in disarray and the

Scottish king was taken to Falaise in Normandy, surrendering the independence of his country to Henry II of England. It was the last attempt by a Scottish king to conquer northern England.

A few centuries later, a thick haar greeted Mary, Queen of Scots when she returned from France to take the throne of Scotland on 19 August 1561. She arrived at the port of Leith, Edinburgh, and as her French navy ship sailed in, 'Borne up the Firth of Forth on a fresh east wind, the fog settled for miles along the shoreline, heavy and impenetrable,' reported a chronicle. Mary's ship remained unseen by the locals, and she landed without any reception, quite possibly an ill omen of the grim political and religious situation facing her in Scotland, and which eventually led to her tragic end. As John Knox, leader of the Protestant Reformation, observed the day she landed, 'The mist was so thick and so dark... The very face of heaven, the time of her arrival, did manifestly speak what comfort was brought into this country with her, to wit, pain, darkness, and all impiety.'

Sea fog can strike all around the British coastline, most famously encapsulated in *The Times* headline 'Fog In Channel – Continent Cut Off'.

Spanish Armada

The Protestant Wind

In the late 1500s, relations between Catholic Spain and Protestant England reached breaking point. The English were harrying Spanish ships in the West Indies and supporting rebels in the Spanish-held Netherlands. The Spanish wanted a Catholic monarch to rule in England. Eventually, King Philip II of Spain decided to use a mighty armada of 130 ships carrying 30,000 men to defeat the English navy and then link up with his forces in the Spanish Netherlands to invade England and defeat Elizabeth I.

By 1588 the Armada set sail from Lisbon, the largest fleet ever seen in Europe and seemingly invincible. By late July it arrived in the English Channel and far outnumbered the smaller English fleet docked in Plymouth. But that July was cold and stormy around the British Isles, as the Duke of Medina Sidonia, commander of the Armada, wrote on 27 July as they sailed past Lizard Point, Cornwall. 'It blew a full gale with very heavy rain squalls and the sea was so heavy that all the sailors agreed that they had never seen its equal in July, Not only did the waves mount to the skies but some sea broke clean over the ships.'

Battles broke out in the Channel between the Armada and the English, but the winds chopped and changed direction as depressions blew through followed by high pressure and settled weather. The Spanish held their tight formation of ships, but could not find safe anchorage and were driven on by westerly or southwesterly winds until they anchored off Calais on 6 August. There the English waited until night-time before using the wind and a spring high tide to launch eight fireships set ablaze to strike at the heart of the Armada. The sight of the flaming ships, exploding with gunpowder and shrapnel, sent the Spanish into pandemonium. They slashed their mooring lines and set off in such a panic the ships collided with each other. Next day they found themselves blown on fresh winds further up the coast towards the treacherous Zeeland sandbanks off the Flemish coast where they faced running aground, whilst the English chased them down. Several hours of fierce broadsides from Drake's ships shattered the Spanish galleons before heavy weather brought the battle to a halt. Those winds drove the Spanish ships ever closer to the shore where the entire fleet faced being wiped out, before the wind suddenly changed direction again and blew the stricken Spanish ships northwards, in what seemed like a miraculous escape. In fact, they faced new dangers as they were battered by a North Sea storm.

By now the Armada was in no condition to turn back and fight their way back through the Channel and in any case they were blown further north. So they decided to sail back to Spain around Scotland and Ireland. But once they rounded Scotland the ships faced the full brunt of more Atlantic storms. Many of the ships were badly damaged, supplies were running low, and the men were becoming seriously ill. By far the worst tempest struck on 21 September, as one ship reported: 'It began to blow with most terrible fury. There sprung up so great a storm on our beam [westerly] with a sea up to the heavens so that the cables could not hold nor the sails serve us and we were driven ashore with all three ships upon a beach covered with fine sand, shut in on one side and the other by great rocks.' The storm took a heavy toll of 17 ships lost and was so severe it may have been the remains of an Atlantic hurricane, possibly similar in force to ex-Hurricane Debbie that smashed into the west coast of Ireland with devastating force on 16 September 1961.

Far more Armada ships were wrecked by the storms than inflicted by the English. Only around a half of the original Armada made it back to Spain and the defeat shattered the image of Spanish invincibility. The victory saved England from invasion, broke the overseas Spanish-Portuguese domination of the seas and opened the way for England's colonial expansion. And Elizabeth I and her navy became a force to reckon with.

Timely Thunderstorms

Battle of Marston Moor, 1644

In olden days, battles tended to be fought in the summer, when troops could be moved and provided for in reasonable conditions, but even so they still ran the risk of violent weather. In the English Civil War, between the Royalists and Parliamentarians, the Royalist commander Prince Rupert was moving south to meet up with the King, having lifted a siege of his fellow forces at York. But the Parliamentarians gave chase, and the two sides met on 2 July 1644 at Long Marston, a flat moorland not far from York. At first the opposing forces eyed each other up in a standoff that lasted so many hours that Prince Rupert stood down his troops in the evening so they could eat, convinced that the enemy would not attack until next morning. But at around 7pm, a thunderstorm erupted, and seeing their opportunity to seize the initiative, the Parliamentarians launched a lightning attack. The Royalists were caught by surprise and all seemed lost as they were heavily outnumbered. However, they were not overwhelmed and, in a fierce fight, the battle ebbed and flowed between the two sides. But the deciding factor was a valliant assault by one of the Parliamentarian leaders, Oliver Cromwell, who, despite being wounded, swept his troops in a wide arc across the moor, defeating all that stood before them and won the battle.

Long Marston is believed to be one of the largest battles on British soil. Around 4,000 Royalist troops were killed against only a few hundred of the Parliamentarian forces. Prince Rupert's reputation as an all-conquering commander was severely damaged, whilst Oliver Cromwell's reputation soared. The battle decided control of the north of Britain and the royalist Northern army was effectively destroyed.

Another thunderstorm also came to the rescue of the British two centuries later, at the Battle of Waterloo, in 1815. The Duke of Wellington faced the overwhelming might of Napoleon's forces in Belgium. Napoleon was desperate to crush Wellington before the Prussian army led by Prince Blücher could reinforce the British. But on the night of 17 June, an intense thunderstorm turned the fields of Waterloo to mud. 'The water ran in streams,' described one English soldier. 'We were as wet as if we had been plunged over head in a river.'

Napoleon planned to attack at around 8am the following day, but his artillery commander feared the muddy conditions made it impossible to move guns quickly. So the attack was delayed, and when it was launched the quagmire slowed down the cavalry, men sank ankle-deep in mud, the infantry had to wade through fields of wet rye that left their gunpowder wet, cannon fire sank harmlessly in the boggy conditions and it was difficult to see the enemy through mist and gun smoke.

However, the wet conditions were a gift for Wellington to stay fixed in defensive positions and hang on for the arrival of the Prussian reinforcements. But as the ground dried out during the day, Wellington was left exposed and by mid-afternoon was facing defeat just before the Prussians arrived at 4pm. The allies decisively routed the French but, as Wellington later admitted, 'It was a close run thing.' The future of Europe was secured for a hundred years – thanks in no small part to the weather.

Lightning Disaster
Athlone 1697

The searing heat of lightning is five times hotter than the surface of the Sun, and is notorious for setting off fires. Possibly the worst recorded lightning disaster in history struck on 27 October 1697 in Athlone, in the heart of Ireland. A brutal thunderstorm broke out during the small hours, unleashing a violent wind that demolished the guardhouse of Athlone Castle, followed by torrential rain falling so loudly it sounded like stones cascading down on the cobbled streets. Finally thunder and lightning ripped through the sky in a continual blaze:

'Heaven and Earth seemed to be United by the Flame of this Lightning', exclaimed one report of the time.

Following three ear-splitting claps of thunder, a bolt of lightning ripped into the Castle, striking deep into the arsenal and blowing up 260 barrels of gunpowder, 1,000 hand grenades, various incendiaries, 220 barrels of musket balls and pistol balls, as well as great stores of pick-axes and other ironmongery. This vast bomb blew up in such a spectacular explosion that it demolished the Castle and set fire to the surrounding town, completely destroying 64 houses and burning almost all the rest. The damage was so severe that the entire castle and town had to be rebuilt, although only eight people were recorded as killed and about 36 injured, as one sobering account added: 'None Killed of Note'.

But the disaster is still shrouded in some mystery. Could it have been caused by a fireball rather than a bolt of lightning? Apart from lightning, another description included an intriguing phenomenon: 'Just above the Castle, and at the last of three Claps, in the Twinkling of an Eye, fell a Wonderful and Great Body of Fire, in figure Round... [which fell] down the Magazine took Fire and blew up the Granadoes.' Exactly what this fireball was remains very uncertain, but is reminiscent of a violent incident of ball lightning.

Another lightning disaster struck South Wales on 24 July 1994. An intense thunderstorm swept unleashed lightning that hit the huge oil refinery at Milford Haven, Pembrokeshire at around 8am. Power controls across the site were sent into turmoil, setting off false alarms at the rate of one every 2–3 seconds in scenes so chaotic that it became impossible to read them, let alone act on the information. The crisis carried on for five hours, during which the false alarms disguised a catastrophic malfunction on one crucial valve. At 1.23pm, the valve was stuck and set off 20 tonnes of liquid hydrocarbon bursting through a pipe, which exploded into a fireball with a force equivalent to 4 tonnes of high explosive. Fires were set off all over the plant, flames shooting 100ft (30m) high and shops in nearby Milford Haven town about 2 miles (3km) away had their windows blown in. Had it not been a Sunday the casualties would have been far worse than the 26 people who only sustained minor injuries. It took over two days to put the fires out, rebuilding costs were estimated at £48 million and loss of production was even greater.

Another lightning disaster struck as a 'bolt from blue'. On 9 July 1984, part of York Minster was set ablaze by lighting from what seemed to be a clear sky at night. A thunderbolt blasted the roof of the South Transept, which burst into flames. More than one churchman afterwards wondered if the fire was divine retribution, just three days after the Rt Rev. David Jenkins, the Bishop of Durham, had made controversial remarks about Christianity. The restoration of the Minster took over four years and cost more than £1 million.

One theory is that the fire was caused by a positive lightning strike. Normally, a thundercloud is full of negative charges at its base and positive charges at the cloud top. During the storm, negatively-charged lightning shoots out from the cloud down to ground, but as the storm dies out, its negative charges become exhausted. This leaves an overhang of positive charges left at the top of the cloud, and these can sometimes go to earth in one spectacular 'thunderbolt', an ear-splitting lightning strike. They can also shoot horizontally more than 10 miles (15km) away, where there may be little sign of any thunderclouds, so they seem to come out of the blue. Positive lightning is far more powerful than negative strikes, possibly reaching 300,000 amps and 1 billion volts. This makes them especially dangerous, and is thought to set off large numbers of forest fires and cause power-line damage worldwide.

Waterspout Fury

On 17 June 1760, Methodist leader John Wesley recorded an account of a strange happening from an eyewitness in Cornwall. 'A round pillar, narrowest at bottom, of a whitish colour, rose out of the sea near Mousehole and reached the clouds. It dragged with it abundance of sand and pebbles from the shore; and then went over the land, carrying with it corn, furze, or whatever it found in its way.' One woman riding along the coast was thrown from her horse after the whirling funnel came close by.

It was a waterspout, a great funnel of water drawn up from the sea into a cloud above, looking rather like a giant elephant's trunk. These can be either a tornado passing over water, and be quite violent, or they come from a column of warm air rising off a warm sea into humid air above.

A waterspout in September 1886 caused even more pandemonium, when it travelled across the sea and struck the coast at Kelvey Hill, Swansea, with extraordinary force releasing 'great volumes of seawater'. The torrents of water rushed down the steep slopes, bursting through a row of houses, sweeping men, women and children off their feet and even out of their homes, some of whom were found buried in debris and mud. But even though several people were injured, none were killed. Forty families were made homeless and some 8,000 tons of earth and huge boulders dumped at the foot of the hill, blocking the local railway line. Hundreds of navvies were brought in to clear the wreckage.

Waterspouts can also inflict serious damage in mid-air. In February 2002, a hair-raising incident occurred when a helicopter carrying oil workers from a North Sea oil rig was struck on its tail rotor by a giant waterspout off the Shetland Islands. The Super Puma helicopter, with 18 people on board, was flying at 150m (500ft) when it was hit, and the helicopter 'violently pitched, rolled and yawed'. The helicopter's wild gyrations left a 15cm (6in) gash after all five of the tail rotors touched the tail pylon. Incredibly, the helicopter landed safely without injuries.

Cold Summers And Revolution

Crop failures of the 1790s

Britain was facing turmoil at the end of the 18th century. The American War of Independence had been lost, George III was losing his mind, and Revolutionary France threatened to overrun Europe. Adding to this chaotic milieu, the climate rapidly turned atrocious and left grave shortages of food.

The warning signs had already devastated France, where harvests were ruined by cold wet weather over several years and stoked the flames of revolution. In 1788 a severe drought was followed by a long, severe winter, decimating the wheat crop and leading to even higher high bread prices. Thomas Jefferson, American ambassador to Paris at the time, wrote about that terrible winter: 'So great, indeed, was the scarcity of bread, that, from the highest to the lowest citizen, the bakers were permitted to deal but a scanty allowance per head.' The soaring price of bread sparked riots that became increasingly violent and spiralled out of controlled, fuelling the French Revolution of 1789.

In Britain, the nation was no longer self-sufficient in grain as the population boomed and the industrial revolution led to a huge migration of people from the countryside to overcrowded cities. The deficiencies in food could only be made up by imports, but bad harvests and a deteriorating political situation abroad meant the supply of imported food began to dry up. To make matters worse, the home-grown harvests suffered as well.

The rot began to set in during the winter and summer of 1792, which was so wet and cold that the staple diet of oats in the uplands of Scotland was threatened. In one parish, Lauder, 'it was the end of December before the harvest was finished after a greater part of the crop was destroyed by frost and snow. None of the farmers could pay their rents and many went bankrupt.'

The poor harvest led to shortages of foods, prices rocketed and the government prohibited the export of grain. Food riots broke out across the country, and in Swansea hundreds of people raided farmhouses and even took farmers hostage for food.

The situation deteriorated even further in 1793 when revolutionary France declared war on Britain and in return the British government ordered a naval blockade of France. The following years produced more misery. The winter of 1795 was hit by cold followed by floods, which inevitably wrecked the harvests. A severe shortage of wheat triggered soaring food prices, workers were laid off and wages cut. Waves of food riots spread across the entire country as workers seized food stocks and took over markets, forcing traders to lower their prices. In Devon alone there were 43 food riots.

The food shortage at home and the success of the French armies on the continent produced a strong backlash against the war itself and in an atmosphere of political crisis, the prime minister, William Pitt the Younger, agreed to recall parliament and discuss the possibility of peace negotiations. On 29 October, a crowd estimated at 200,000 surrounded Westminster. The king was hissed and booed, his carriage stoned and the crowd demanded bread and peace. The government responded with a proclamation against seditious assemblies, and speeches 'inciting people to hatred or contempt' of the king or the government became punishable by death. The government was also forced to resort to stringent emergency measures, including bans on export of grain, and even the manufacture of hand and hair powder from flour and the use of grain for malt distilling were prohibited.

Paranoia reached new heights when a French invasion fleet almost landed at Bantry Bay, Ireland, in December 1796, but were defeated by a severe storm that blew the fleet out to sea. The French returned in February 1797 and landed at

Fishguard, west Wales, although the invasion was thrown into disarray and surrendered.

By 1797 Britain was almost on the verge of bankruptcy, and fearing further enemy landings, the government made plans for a withdrawal from the coast using a scorched-earth policy. Supplies of home-grown food were now even more crucial and the government called for agricultural surveys to discover precisely how much land was used for farming in order to boost production.

Another cold wet summer in 1799 was followed the next year by the complete opposite – heat and drought. In July 1800, there was virtually no rain at all and forest fires broke out. The drought was followed by incessant rains, and the crops failed again. In Worcester one report summed up the disaster: 'A scarcity of bread having been experienced for some days, the populace very erroneously considered the bakers to have been the sole cause of it and therefore vented their fury upon the promises of the latter. ... the scarcity of bread proceeded from a cause no-one could control. This was the stopping of many of the corn mills for want of water after the long drought of this summer.'

Between April 1799 and July 1800, the cost of wheat tripled and across the country the rioting grew even more furious. Revolution was in the air and a poster in Bath demanded 'Peace and Large Bread or a King without a Head,' and another, 'Bread or Blood... Have not Frenchman Shewn You a Pattern to Fight for Liberty?' In Birmingham and Nottingham rioters broke into food stores and the army was called in to put the riots down. Bread riots brought London to its knees, and shops were smashed up. In Nottingham 'Granaries were broken into at the canal wharfs, and it was really distressing to see with what famine-impelled eagerness many a mother bore away corn in her apron to feed her offspring.' Troops were used to put down the riots.

The government banned the distillation of spirits from grain, and even the powdering of wigs with flour was banned. Soup kitchens were set up, run by volunteers with private donations. The government urged the growing of potatoes and the improvement of agriculture.

The country desperately sought imports. A large part of the nation's grain came from Prussia, Poland and Russia. But the Baltic ports through which the grain was shipped were effectively under the control of the volatile Tsar Paul I. As Russia, Denmark, Prussia and Sweden formed an 'Armed Neutrality', Tsar Paul imposed an embargo on British ships and imprisoned some 4,000 crew and sequestered all British property in his empire. Relations between Britain and Russia rapidly sank.

With real fears of a revolution at home, the British government embarked on a desperate mission. The Cabinet decided to send a fleet to the Baltic, break up the Armed Neutrality by force and smash the trade embargo to get much sought-after grain. In April 1801 a British fleet defeated the Danish navy at the Battle of Copenhagen, and were all set to push further up the Baltic when events suddenly defused the situation. Tsar Paul died soon after the defeat of the Danes and the Armed Neutrality broke up, helping to ease the crisis in the grain imports. Even further semblance of normality returned in 1802 with the first decent harvest in years, which alleviated the food shortages further. The food crisis eased, but the Napoleonic Wars took a terrible toll on the whole of Europe.

Storms of Ice

Hailstorms 1808 and 1843

Searing heatwaves are not a new phenomenon of global warming. The heat of July 1808 was so intense that many people died from the searing temperatures, as listed in one edition of the *Gentleman's Magazine*:

> 'A man at Corby in Lincolnshire, on Wednesday, while hoeing thistles, was so overcome by the heat that he died in the field. A woman employed in making hay in a field adjoining the town of Huntingdon complained of being ill from the heat, and died before she could get home. A woman at Billesdon in Leicestershire on Wednesday died of the excessive heat; as did two others at Houghton and Ansley and two or three also in Leicester.'

This was clearly no ordinary summer. Temperatures may have broken the 40°C (105°F) level, which would be even hotter than the official highest record of 38.5°C (101.3°F) in 2003 (see 'Summer Heat 2003'), but there were few reliable weather observations 200 years ago to be sure.

Eventually, on St Swithin's Day, 15 July 1808 the heat finally erupted into stupendous thunderstorms. 'At Gloucester a fire ball burst about 11 o'clock in the College-Green carrying away one of the pinnacles upon the west end of the cathedral,' wrote one observer. The storms also produced a brutal hailstorm that swept the southwest, tearing a path between Bath and Bristol. These were no ordinary hailstones, though. There were curious reports from Gloucestershire of ice that looked like fragments of plates smashed up 'broad, flat and ragged', described the *Bath Chronicle*. Some hailstones at Templecombe in Somerset were reported to be over a foot long (30cm), more like small bombs falling from the sky than hailstones.

At Mells manor, home of Colonel Horner, of the 'Little Jack Horner' nursery rhyme whose 'plum' was the manor at Mells, wrote that over 3,000 panes of glass were broken in his house, hothouses and garden. Rooks, pigeons and pheasants lay dead on the ground, and crops were beaten down. People had to run for their lives to avoid being stoned to death, and north of Mells 'A farmers' boy... was so battered by the hail that he was black and blue.'

Hail over 4in (10cm) across fell at Batcombe and was even larger in some other places. Trees were stripped bare, cereal crops flatted as if 'a heavy roller had repeated passed over', and in an area famous for its fruit, 'apples sufficient to make forty hogsheads of cider destroyed,' noted Mr Melhuish of Suddon House, near Wincanton. Everywhere was covered in debris from the onslaught.

If a storm of such intensity struck a built-up area today, the damage would be horrendous and casualties horrific. Fortunately the worst of the hailstorm of 1808 managed to avoid any towns by taking a path between Bristol and Bath. In fact, the isolated nature of the hailstorm meant that it passed off almost unnoticed in other parts of the country.

However, another brutal hailstorm, on 9 August 1843, swept across the Midlands and East Anglia and inflicted huge damage. The storm had all the hallmarks of a supercell, a type of thunderstorm notorious for huge damage in 'Tornado Alley' in the US Midwest. Not only are supercells extremely powerful, but they also last far longer than most ordinary thunderstorms.

In 1843, the storm began as ink black thunderclouds spat lightning and rocked with thunder, tornadoes tore through the countryside, rain fell in spectacular cascades and hail fell in such vast fusilades that the ice lay up to about 5ft (1.5m) deep in places, with hailstones up to 10in (25cm) across. From Oxfordshire to Norfolk, the hailstorm pulverised roofs, smashed windows to pieces and blocked roads in deep piles of hailstones. Fields were pulverised and sheep stoned to death. In Chipping Norton, Oxfordshire: 'Who can describe the smashing of windows in the town, the rattling of the broken panes as they fell among various articles, or were driven across various rooms,' a correspondent wrote in *Jacksons Oxford Journal*. The *Banbury Guardian* reported 50,000 panes of glass were broken during a 20-minute blitz. Hailstones described up to 9in (23cm) smashed through roof slates. There was a report of flat discs of ice more than 1ft (30cm) across falling amongst the hailstones, reminiscent of the peculiar ice 'bombs' in the 1808 storm. There were two incidents of kettles over fires being shot through by hailstones as they fired down chimneys like bullets. But at least one person found a use for the hail, when Mr Cartwright of Aynho Mansion, Northamptonshire, used the heaps of ice to fill his ice-houses and cool his wine.

Throughout the great length of the storm's track, any window facing northwards was smashed. Greenhouses were demolished, and ancient church windows destroyed. At Wimpole near Cambridge: 'The destruction of property was dreadful! All the windows on the north side of the Mansion [Wimpole Hall] were broken, all the hothouses, and every window facing the north in many of the cottages! The land before it was as the Garden of Eden, behind it a barren and desolate wilderness'.

In Cambridge, the wreckage included colleges, county courts, the university press and botanic gardens, even ripping out leaded window framework. 'Scarcely a pane of glass was left whole in the town,' reported *The Times*.

In Norwich, stalls in the market place were swept away and a performance at the Theatre Royal was stopped as the cast and audience panicked at the deafening roar of the storm outside. Cellars and basements across the city were flooded with ice and rain, and at one pub the flood loosened the earth under the beer cellars and caused it to collapse into a great hole, along with beer barrels.

Across large parts of Norfolk crops were so badly damaged that the local council levied a special voluntary relief fund to compensate hundreds of farmers who were left destitute as their buildings and crops were left in ruins. Hail insurance was available at the time from The Farmers & General Fire & Life Insurance & Loan & Annuity Company, founded in 1840, but its losses from the 1843 storm were so severe that the shareholders were asked to provide additional capital to help meet the claims. Shortly afterwards the company was merged with another insurer, which eventually grew through more amalgamations to become the Norwich Union, one of the UK's largest insurance companies.

None of the hailstones in 1808 and 1848 were officially measured, so have not made it into the record books. But it is clear they were far larger than the UK's official record for the largest hailstones of 3.15in (80mm) in diameter on 5 September 1958 at Horsham, West Sussex.

The Year Without Summer
1816

In April 1815, Mount Tambora in Indonesia exploded so violently a third of the volcano vanished in rocks and dust. It was possibly the biggest eruption in recorded history, killing around 10,000 people from the eruption and many more from starvation and disease from the devastated landscape left behind. Ash and sulphuric acid was blasted so high it penetrated 27 miles (43km) high into the stratosphere and then spread around the globe, hanging as a veil over the sky and blocking out sunlight for years. As a result, global temperatures dropped so drastically that the following year in the northern hemisphere became known as the year without a summer.

In Britain the summer of 1816 was wet, cold, and utterly wretched. The foul weather rotted crops and led to serious shortages of food. Farmworkers were left unemployed, grain prices soared and mobs went on the rampage for food – in one riot, some 2,000 people in Dundee ransacked over a hundred shops and a grain store. In Ireland, rain fell on 142 out of 153 days of summer, potato crops rotted and an estimated 60,000 people died of famine or typhoid.

Across Europe there was desolation, at a time when the continent was emerging from the chaos of the Napoleonic Wars. In the grim climate of 1816, crop failures in France led to riots that shook the new constitutional monarchy of Louis XVIII and Tallyrand. Starving Germans baked straw and sawdust into loaves of 'bread', and in Switzerland people ate moss in desperation. At least 200,000 people died from famine in Europe, and the atrocious weather was also blamed for a typhus epidemic from 1816–19 that killed millions more. Thousands of people gave up and emigrated to the USA only to find conditions there were just as bad – the northeast suffered snow in June and frosts in July that ruined harvests and, in turn, drove a mass migration of farmers westwards across the prairies.

The shocking weather plagued much of the northern hemisphere. The Indian monsoon failed and the resulting famine triggered a major outbreak of cholera that spread into what became the world's first pandemic of cholera. Cold weather killed trees, rice crops and water buffalo herds in northern China and disrupted the monsoon season there as well.

But the dismal weather led to some unforeseen consequences. In Switzerland, Lord Byron rented a villa near Lake Geneva and among his guests were 18-year-old Mary Shelley and her husband Percy. The weather kept the house party trapped indoors, as Mary explained: 'At first we spent our pleasant hours upon the lake, or wandering on its shores. But it proved a wet and uncongenial summer, and the incessant rain often confined us to the house for days on end. "We will each write a ghost story," said Lord Byron.' The result was Mary Shelley's *Frankenstein*, the story of how Baron Victor Frankenstein gave the spark of life to his monstrous being, published in 1818.

That same summer party inspired another masterpiece, by Dr John Polidori, Byron's personal physician, who wrote *The Vampyre*, which eventually inspired Bram Stoker's work *Dracula*. And Byron himself became so depressed in the gloomy weather he wrote 'Darkness', a poem about the extinction of the sun:

'I had a dream, which was not all a dream.
The bright sun was extinguish'd, and the stars
Did wander darkling in the eternal space,
Rayless, and pathless, and the icy earth
Swung blind and blackening in the moonless air;
Morn came and went – and came, and brought no day...'

The volcanic ash of Tambora also created brilliantly coloured sunsets, twilights glowing orange or red near the horizon, purple to pink above as sunlight scattered off the dust floating in the stratosphere. Those vivid skies became a trademark of the painter William Turner, and the fantastic colours he painted may well have been more than just artistic licence. Turner lived through several other major eruptions, including the Babuyan Claro volcano in the Philippines in 1831 that resulted in spectacular sunsets and twilights, as well as green and blue coloured suns.

The year without summer also led to another unexpected benefit. In Germany the soaring price of oats caused a crisis in horse feed that spurred Karl Drais to invent a replacement for the horse – the 'draisine' or velocipede, a forerunner of the bicycle. Drais first demonstrated his velocipede in June 1817, and although the machine proved to be popular to begin with, most cities ended up banning them as a menace to pedestrians.

The Big Stink

1858

London in the mid-1800s was at bursting point. In the last 50 years the population of London had doubled to 2,362,000 and sewage became an enormous problem. Ironically, the sewage problem came from technological progress when the WC was invented. Sewage used to be collected in cesspools and transported out of London, but WCs flushed raw sewage into rivers and streams and finished up in the Thames, where it was trapped by the river's tidal flow. The river soon became the world's largest cesspool.

In June 1858 a scorching heatwave brought the capital to crisis point. The sewage in the Thames gave off the most appalling stench, which became known as 'The Great Stink.' The law courts were brought to a halt and were almost transferred outside London. In Parliament, MPs in the newly built House of Commons clutched handkerchiefs to their noses, and sheets soaked in chloride of lime were hung from the windows of the building in vain attempts to get rid of the smell. The *City Press* newspaper on 19 June 1858 was rather more succinct: 'Gentility of speech is at an end – it stinks; and whoso once inhales the stink can never forget it and can count himself lucky if he live to remember it.'

This was more than just a bad smell, though. Drinking water was taken from the river and London saw 18,000 deaths in 1849 from cholera and a further 20,000 in 1854, caused by the contaminated water. It was thought the diseases were carried by the stink as a 'miasma' – and, by sheer serendipity, when the stink was eventually got rid of, the diseases disappeared as well. It took decades more before it was fully realised that the contaminated water caused disease, and the big stink itself was just a nuisance.

For years Parliament had debated the stinking problem over and over again, and although laws were passed to get a new sewer system, the legislation was too weak to get anything done. Some 137 different schemes had been proposed to rid the Thames of sewage, and all were turned down. But The Great Stink changed all that. 'Parliament was all but compelled to legislate upon the great London nuisance by the force of sheer stench,' *The Times* thundered. With indecent haste, a new law was passed in only 18 days to get a new sewer system, a phenomenal undertaking costed at £3m. It's reassuring to know that even in those days it ended up over-budget, at £20m – around £1.5b in today's money, and dwarfed even Brunel's Great Western Railway, which cost £8m.

A brilliant engineer, Joseph Bazalgette, designed a massive new sewer system to solve the problem. It was an amazing feat of engineering, the biggest public construction project of its time, that intercepted the city's sewage before it reached the river and diverted it downriver where the tide could take it away. Huge steam engine pumps housed in great gothic-looking palaces had to be built to raise the sewage high enough so that it could flow downriver by gravity. A sewage super-highway had to be built, consisting of 82 miles (132km) of megadrains so big that land had to be reclaimed from the muddy banks of the Thames and turned into embankments – what became the Victoria, Albert and Chelsea Embankments. And hundreds of kilometres of new underground street sewers were also built to collect London's waste more efficiently.

Nothing like it had ever been done before, and it became a model for industrialised cities all over the world. The entire project took 16 years to become fully operational, and it made Bazelgette a hero. The stink ended and so did the cholera epidemics.

Today, the Thames suffers another sort of Big Stink. Most of London's sewage still flows through Bazalgette's tunnels. But now the population of the capital is four times larger, the capital is far more built up, and the rainfalls are growing heavier as the climate changes. Even a light shower can overload the combined drain and sewer system, forcing raw sewage into the Thames, polluting the water and killing fish. But now virtually every year exceptionally violent downpours of monsoon proportions are striking London. On 3 August 2004, a huge downpour of rain forced 600,000 tonnes of untreated sewage and urban run-off into the Thames, followed by more storms that added even more sewage. The contaminated water not only starved fish of oxygen and killed thousands of them, but also caused a shocking rise in disease microbes such as Salmonella, Novovirus, and Enteroviruses in the river. Water sports enthusiasts using the river were warned to wash their hands after using the river, as if they had been to the toilet.

After years of prevarication, a new £2bn super-sewer is planned to get rid of these sewage outflows by collecting the rainfall overflows in a new tunnel system that will dump the waste downriver.

The Hot Summer

The summer of 1911 was spectacular, a last Edwardian swan-song before the outbreak of the First World War. Blistering heat stretched from May to September and a new July record of 36°C (96.8°F) was set on the 22nd at Epsom, Surrey. It was also the sunniest month on record, with 384 hours of sunshine logged at Eastbourne and Hastings, a record that still stands.

Whilst the upper classes enjoyed summer parties and country retreats, workers across the country were boiling with resentment. Wages and living conditions were growing worse and seamen, dockers, miners and railwaymen went on strike. In the long, hot summer, tempers flared up and riots broke out. The country was brought to a near standstill, and the government, believing the stability of the country was at stake, sent in troops to quell the troubles. The heat also led to thousands of deaths from the heat and gastroenteritis diseases caused by contaminated waters and poor public hygiene.

The searing heat and drought carried on well into September, and left harvests in a dire state, with vegetables shrivelled up and pasture grass parched brown. 'The cabbages are tough enough to make carpets,' remarked one vegetable buyer at Covent Garden market. Reservoirs ran completely dry, streams withered away and water supplies were turned off. Many northern woollen mills closed down as their streams dried up, throwing thousands of workers out of work. Palls of smoke hung over the countryside from forest and heath fires. But the seaside resorts did a roaring trade, even in September:

'Bathing was freely indulged in at Eastbourne' reported the *Daily Telegraph* on 9 September. *The Spectator* was more concerned by bathing in the Serpentine in Hyde Park 'at unwonted hours', and deploring that 'it would probably see the practice of mixed bathing imported from the coast to our metropolitan waters.'

At Dover on 6 September, under a warm blue sky, Yorkshireman Thomas Burgess, aged 37, started his 16th attempt to swim the English Channel. He was stark naked except for a pair of motorist's goggles and a rubber bathing cap, and was smothered in lard. Badly seasick and stung by jellyfish, his spirits were lifted by his accompanying boat crew singing to him whilst he was kept fortified by hot chocolate, grapes and 20 drops of champagne each hour. He landed near Sangatte, almost an entire day later. Remarkably he had only done 18 hours' training for the swim that year.

Newspapers also carried reports of a crisis in Morocco with Germany and France, and war was mentioned. The long hot summer and autumn finally broke in mid-September and the second half of the month was cool and unsettled.

Rain Records

1912, 1903

August 1912 broke the records for the coldest, dullest and wettest August in history. To cap it all, it was also the rainiest summer on record. Amazingly, this hideous weather came after the astonishing heat of the summer of 1911. But the summer of 1912 was dogged by gales, thunderstorms and cloudbursts of exceptional rainfalls. Nerves were already strained by the huge military build-up in Germany, and army manoeuvres on Salisbury became bogged down on waterlogged ground, an omen of things to come on the Western Front.

At the end of August, rain had been falling for several days before a gigantic deluge on 25 and 26 August. Over 7in (180mm) rain fell in a single day in Norwich and the city collapsed into chaos. Large trees were blown down in the winds, rivers burst their banks, drains clogged up with debris and the streets turned into rivers surging through the city.

In those days, Norwich was densely populated with 100,000 inhabitants living within a mile of the city centre, and thousands of people were trapped in their homes. 'It was pitiable to see these poor people struggling with their adversity, and to think of the cheerless night they must spend amid their sodden surroundings,' the *Eastern Daily Press* wrote. 'The pressure of the water in the sewers burst off the lids of manholes in some instances, and huge fountains of water rose up from them and added to the raging floods around.'

Heroic rescue operations were made in rowing boats, but four people died. About 15,000 people suffered damage to homes and 2,000 were left homeless. 'Not within living memory has there been such an August as that now drawing to a close, nor, in view of this week's terrible experiences one so disastrous,' read one report.

Norfolk was cut off from the rest of the country as 52 bridges collapsed and roads and railways vanished under floodwaters. There was enormous damage to crops and the 'new-fangled' tractors proved useless in waterlogged fields. The rest of that year remained thoroughly wet, and a large area of East Anglia stayed under water through the winter. 'Enormous damage to crops, both of hay and corn, was reported and great quantities of hay were carried bodily away and washed into the sea. A very large area of land remained under water for the whole of the following winter,' reported the *British Rainfall Journal* for 1912.

The summer of 1903 was another travesty, and June was desperately wet and cold. Rain fell continuously at Camden Square, London for 58.5 hours from 13 to 15 June, the longest period of continuous rain recorded in England. At the same time, snows blanketed the Scottish Highlands and even parts of lowland England; *The Times* described conditions around Newmarket, Berkshire: 'The weather was very cold last night, snow and hail accompanying the rain. Much of the hay already cut in the district has been spoiled and fruit crops ruined.'

That summer farmworkers suffered an epidemic of lung disease from mouldy hay and grain.

The interesting feature of both these wet summers is that they can be linked back to volcanic eruptions. The dismal summer of 1912 may have been caused by the huge eruption of Novarupta on the remote Alaska peninsula on 6 June that year, one of the most powerful eruptions of the 20th century. And the rains and cold of 1903 may have been the result of a spate of volcanoes that erupted in Central America and the Caribbean, including the disastrous eruption of Pelee in Martinique, which destroyed the city of St Pierre and killed 29,000 people.

These volcanic eruptions injected clouds of fine ash and acid droplets into the upper atmosphere, screening out the sun and cooling the world's climate for long afterwards. However, no volcanic activity can explain the diabolical summer of 2007, which flooded large parts of Britain (see 'Summer Floods 2007'), and the fundamental cause of which remains unknown.

Lightning Terror At Ascot

The traditional Royal Ascot weather is supposed to be warm and sunny. But on 18 June 1930, on the second day of the races, a thunderstorm broke just as the winning horse crossed the line in the Royal Hunt Cup, and torrential rains flooded the racecourse, leaving the crowds of racegoers running for shelter.

'Terrified Women in Mad Rush' ran a headline in the *Daily Herald*. 'In some of the tents women fainted. Four collapsed together in one small tent. Word sped that a man had been killed by lightning in Tattersall's. The real terror showed then. Men went pale. Women gazed at the storm with fascinated horror, deadly white.'

The victim of the lightning was Walter Holbem, an ex-Sheffield Wednesday football player who had retired as a bookmaker and was working at the races that day. However, the *Daily Herald* was more concerned with the ladies' attire in the storm: 'Women pinned up their frocks in the manner of charwomen and trooped out. It was almost impossible to find their motor cars. Once smart men and women jostled, ran, searched. Those who set off down the tunnel to the station found two feet of water. They went across the field ankle deep in mud. Their clothes? They were past caring for them.'

Twenty-five years later, lightning struck Royal Ascot again. The third day of the races, on 14 July 1955, was the hottest day of the year so far. But the skies grew increasingly dark as black clouds rolled in, and suddenly broke in a ferocious downpour of rain.

Blinding flashes of lightning raked the sky, then one bolt soared over the grandstand and struck metal fencing close to crowds packed inside a marquee. In chaotic scenes, dozens of people were scythed to the ground, 'toppled like dominoes', some even lifted clean off their feet. One victim described the lightning strike 'like being stabbed,' and another said he felt a shock tear through his arm as he was thrown to the ground. The scene afterwards resembled a battlefield, people lying injured, unconscious or wandering around dazed and confused. Fleets of ambulances rushed to the racecourse and ferried 48 casualties to the local hospital, which rapidly became overwhelmed by the scale of the disaster. Two people died and the others were injured with burns and shock.

Across England that same day several other fearsome thunderstorms left five other people killed by lightning, in what was one of the worst single day's thunderstorm casualties on record in the UK.

The Weirdest Showers

When a shower breaks out, the most you expect to be showered with is raindrops or hail. But on 15 July 1841, a heavy thunderstorm over Derby brought a shower of something much stranger, as described in the *Sheffield Patriot* newspaper:

> 'Hundreds of small fishes and frogs in great abundance descended with the torrents of rain. Some of the fish had very hard pointed spikes on their backs, and are commonly called sticklebacks. The frogs were from the size of a horsebean to that of a garden bean; numbers of them came down alive, and jumped away as fast as they could, but the bulk were killed by the fall on the hard pavement. We have seen some alive to-day, which appear to enjoy themselves, in a glass of water.'

This was not a unique incident. On 9 February 1859, the Rev John Griffith, vicar of Aberdare, South Wales wrote to *The Times* with an account of another strange shower from one of his parishioners, John Lewis, a timber sawyer at Nixon's timberyard. 'On putting my hand down my neck I was surprised to find they were little fish. By this time I saw the whole ground covered with them. I took off my hat, the brim or which was full of them. They were jumping all about.' Rev Griffith sent some of the fish to Dr Gray, of the British Museum, who identified them as 'very young minnows,' but added, 'On reading the evidence, it seems to me most probably only a practical joke: that one of Mr Nixon's employees had thrown a pailful of water upon another, who had thought fish in it had fallen from the sky'.

The Times reported another weird shower, when thousands of frogs fell at Trowbridge open air swimming pool in June 1939. 'I turned and was amazed to see hundreds of tiny frogs falling,' said Mr Ettles, superintendent of the swimming pool. 'It was a job to walk on the path without treading on them', remarked one woman.

And on 5 June 1983, the coast of Dorset and Hampshire was hit by a terrifying thunderstorm, with torrential rain and winds so fierce that trees were torn down and dozens of small boats capsized at sea. A barrage of hailstones the size of golf balls smashed thousands of windows and a helicopter nearly crashed after one of its two engines disintegrated after sucking in ice. In Hampshire, a bolt of lightning blasted a car across a road and left it upside down. But the storm also dropped huge pieces of black ice, some of it 3in (7.5cm) across, over a swathe across Bournemouth, Christchurch and Poole. On closer inspection they turned out to be lumps of coal encased in ice. The Bournemouth Meteorological Registrar explained: 'Reports of coke have fallen all over the Bournemouth, Poole and Christchurch area. I investigated several reports and found the pieces to be the same. At one lady's house I picked up 92 pieces of coke.' Further along the coast at Brighton, a man had the shock of his life when a crab 10in (30cm) across and frozen in ice dropped close to him.

These surreal events were probably all caused by tornadoes or waterspouts. Eyewitnesses had seen a waterspout at Brighton on 5 June 1983, which may have sucked up the crab, frozen it in the thundercloud and dumped it on land. Another tornado may have swept through a coal merchant's yard and sucked up a load of coal before showering it far and wide.

If a tornado whips across a lake or pond it can also hoover up small animals and carry them up into the thundercloud above. The tornado soon dies away, but the cloud carries on, eventually grows too heavy and drops its rain and frogs miles away in a bizarre shower.

The Greatest Weather Forecast

D-Day 1944

Never in history has so much turned on a weather forecast as it did on D-Day, the Allied invasion of France in June 1944 during the Second World War. The weather, tides, waves and hours of daylight all had to be right for the greatest seaborne invasion ever attempted in history. Originally, May was the chosen as the ideal month, but the landing craft were not ready and the date was pushed back to 5 June 1944. But after that there were few other viable dates left in the calendar.

All seemed promising when May ended in a scorching heat-wave, so hot it broke the temperature record for the month. It seemed like the weather had settled into an ideal pattern ready for D-Day, and the invasion plans gained momentum. But June roared in with the worst storms known for the time of year as freak gales raged through the English Channel and the invasion plans looked to be scuppered.

The entire invasion plans rested on the advice of the Allied weather forecasters, two independent teams of meteorologists, American and British. Their forecasts were assessed by one senior meteorologist, Dr James Stagg, who made the final decision for the Supreme Commander, General Eisenhower.

But the opinions of the two teams of forecasters were badly divided. The Americans based their forecasts on looking back at past records and seeing if they could match the present weather to any pattern from the past. However, the Channel storms of June 1944 were unprecedented in 40 years of weather observations. Despite this drawback, the Americans confidently predicted good weather for 5 June. The British, on the other hand, calculated the forecast using physics and maths, although in those days it was a struggle to make even a 24-hour prediction. And as D-Day grew nearer the rift between the two groups of forecasters widened.

Stagg felt that the Americans were being wildly optimistic with their forecast, even reckless. At a crucial conference of the Supreme Command on 3 June, Stagg went with the British prediction of storms on 5 June that would wreck the invasion taskforce. At the next meeting at 4.15am the following day, 4 June, Stagg could only report that the outlook looked even bleaker, and Eisenhower was forced to accept that the entire D-Day schedule would have to be postponed 24 hours, leaving one million men waiting in limbo along the English coast, lashed by winds and rain.

Indeed, 5 June was atrocious, with Force 6 gales that would have wrecked the entire invasion. And there seemed to be no break in the foul weather. But then an unexpected ray of hope appeared from an unexpected source. A Royal Navy ship stationed south of Iceland picked up a sustained rise in atmospheric pressure, a sign of more settled weather on the way. Although at this time the Channel was battered by a howling gale, it looked like the storm would blow through quickly and leave a brief breathing space of calmer conditions on 6 June before the next storm rolled in. The window of opportunity would last just long enough and be just on the borderline of acceptable conditions for the air force, navy and army.

There was disbelief from the Allied Command when Stagg made the upbeat forecast. Any hope of calm weather seemed long gone, and prospects were being looked at for another attempt at D-Day two weeks later. But Eisenhower went with the new forecast and announced D-Day would go ahead on 6 June. Within hours an armada of 3,000 landing craft, 2,500 other ships and 500 naval vessels were launched and that night 822 aircraft, carrying parachutists and towing gliders, flew to the Normandy landing zones. The weather conditions were, as predicted, just within the operational limits. In fact, the weather was still rough enough that many of the defending German troops had been stood down.

The forecasters not only predicted D-day with stunning accuracy, but the storm that followed also appeared on time, wrecking some of the Mulberry floating harbours that were towed across the Channel to make a temporary port on the Normandy coast. Had D-Day been postponed on 6 June, the next available date was again wrecked by storms. The D-Day weather forecasts were probably the most momentous weather forecasts in the history of meteorology.

In fact, they were so good that a re-run of the weather conditions leading up to D-Day were made using modern computer forecasting techniques, and the results were found to be remarkably similar to the original forecast.

Flashflood Disaster

Lynmouth 1952

Every summer brings flashfloods – intense floods usually triggered by an exceptionally violent burst of rain that can strike with little warning and often catches people by surprise. A flood of just 6km/h (4mph) flowing water, 15cm (6in) deep, can sweep someone off their feet; at 13km/h (8mph) and a depth of 60cm (2ft), floodwater can easily sweep away a car.

Now picture an avalanche of water over 3.6m (12ft) high, carrying 90 million tons of water, rock and trees, charging down a narrow valley into a small town. That was the scene in one of the most devastating flashfloods on record in 1952 in north Devon. Exmoor and the surrounding countryside had been pounded by heavy rains from thunderstorms throughout August, and the weather was so awful that many holidaymakers cut short their stays and went home or cancelled their bookings altogether. But as the ground became saturated, the rainwater ran off in cascades into the local rivers, which rapidly swelled.

The sleepy seaside town of Lynmouth lies at the edge of Exmoor where the rivers East and West Lyn come together in a steep, narrow valley before passing through Lynmouth into the sea. On 15 August a storm of astonishing proportions deluged some 9in (225mm) of rain in only 21 hours. There were later suggestions that around 1ft (300mm) may have fallen on parts of Exmoor, which would have set a new record rainfall for Britain.

As the rains crashed down, people in Lynmouth watched with growing apprehension as roads were awash with sheets of rainwater that started to pour into basements. Staff at the Lyndale Hotel at the bottom of the town tried to bale out the hotel's cellar with mops and buckets, but it was futile. At around 5.30pm the rains grew to a crescendo of terrifying proportions, and within a couple of hours water 1ft (30cm) deep was surging down the roads, cutting off the Lyndale Hotel from the rest of the village.

Worse still, the steep slopes of the valley funnelled torrents of rainwater that ripped down trees and boulders. As the debris came rushing downhill, it became jammed against bridges and created dams, and when these eventually collapsed under the huge surges of water they unleashed an avalanche of some 90 million tons of floodwater, trees and boulders and ripped through the high street of Lynmouth. Buildings were literally torn apart, roads shredded and bridges collapsed before the debris shot into the sea.

Local fisherman Ken Oxenholme told a BBC filmcrew how he escaped from the harbour, which was washed away with the fishing boats. 'Boulders were coming down and two buildings fell down before our very eyes... We saw a row of cottages near the river fold up like a pack of cards, swept out with the river with the agonising screams of some of the local inhabitants, who I know very well.'

The Lyndale Hotel was battered by the floodwaters. Tom Denham, owner of the hotel described the sheer horror of being imprisoned inside: 'About half-past nine there was a tremendous roar. The West Lyn had broken its banks and pushed against the side of the hotel, bringing with it thousands of tons of rocks and debris in its course. It carried away the chapel opposite and a fruit shop. Three people in the fruit shop were swept against the lounge windows of the hotel. We managed to pull them through a window in the nick of time.'

The water and debris burst open the front door of the hotel and the 60 staff and guests inside rushed upstairs. The thundering waters outside laid siege to the hotel, piling rocks and debris against the south wall of the hotel so high they reached up to the first floor of the hotel, pushing the water level even higher. The onslaught finally brought down the lounge and one side of the hotel.

One guest remembered that horrific night: 'We heard the floor below cave in as boulders broke through a window... the building was shuddering all night. We just sat there and talked, walking around sometimes – how we talked. We all got out at dawn through a broken bedroom window straight onto a pile of boulders.'

The *Western Morning News* described the aftermath at both Lynmoouth and the nearby village of Barbrook: 'Superlatives are too puny to describe the calamity, which has befallen Lynmouth and Barbrook. Deaths on a wartime scale, destruction at Barbrook worse than in the heaviest blitz, hundreds of residents and visitors personally ruined and destitute – the story stuns the human mind'.

Masses of boulders and broken trees were heaped everywhere. An idea of the height of the floodwater came from a watermark along the side of the Lyndale Hotel 55ft (18m) above the normal river level. The whole bay below was littered with the debris of shattered buildings, pieces of furniture, cars, lamp posts, and much else.

The toll of devastation was extraordinary: 420 people made homeless, 36 buildings completely wrecked and 72 others badly damaged, 28 bridges demolished, and 95 cars damaged or wrecked and several hundred trees were washed away. A total of 34 people were killed, although it took weeks to recover all the bodies.

The flood problem was partly man-made. Much of the town was built on the old riverbed of the Lyn, which was diverted into a narrow channel under the main street. As the floodwaters overwhelmed the culvert, the river burst free of its confines and sought out its old course.

Since then, stories have surfaced in the media that something more sinister than natural causes triggered the flood. Experimental rainmaking operations by the RAF were being made around that time, which involved flying aircraft through clouds and spraying them with dry ice or silver iodide to encourage more water droplets to form and make rain. However, there were no such flights anywhere near Lynmouth at that time. And what is even more certain is that the torrential rainfall came from an intense Atlantic depression stalled over the West Country and South Wales. No amount of artificial rainmaking could have made the slightest difference to such a vast weather system.

When the Sahara Fell on England

1968

It sounds like a biblical story, when blood red showers rained down over southern England and the Midlands on 1 July 1968. It was a blistering hot day, with temperatures soaring up to 32°C (90°F) in London. Tower Bridge was so hot it got jammed and stopped the bridge closing, causing horrendous traffic jams in the evening rush hour. And at Wimbledon tennis a record 550 people needed medical attention from the heat. But as a cold front swept through a remarkable shower of blood-red rain fell, with some reports of yellow or pink rainfalls as well.

Afterwards everything outside was coated in a thin layer of red, orange and yellow sand. It looked like the seaside had dropped from the sky. Washing was ruined, plants were choked and Water Boards reported a big surge in water demand as people washed their cars clean. Hundreds of alarmed callers rang the police, newspapers and councils asking what had happened – was it a health hazard, pollution, nuclear explosions or vandalism? More coloured showers of foreign dust struck the following day. The weather forecasters could explain it, though, as sand from the Sahara, swept up high into the atmosphere by sandstorms and blown over North Africa, Spain and eventually washed down in rain over England.

In fact, the Sahara is the biggest dust bowl in the world. Storms there sweep up to 1 billion tonnes of dust into the atmosphere each year, and some of that sand reaches Britain about three times a year on average, although it is usually washed away in the rain and hardly gets noticed.

However, there may be a more sinister side to Saharan sand showers. On 13 February 2002, a thin veil of Saharan dust arrived over Britain and some of it may have washed down over the northeast of England. Just six days later, on 19 February, the first outbreak of foot and mouth disease appeared amongst cattle at Heddon-on-the-Wall in Northumbria, the first outbreak of the disease in Britain for 34 years.

The interval between the dust shower and the outbreak was about the time it takes for the virus to incubate, and, as microbiologist Dr Dale Griffin commented, 'that's one heck of a coincidence.' Satellite pictures clearly showed that the most intense part of the dust shower fell over northeast England. But the idea that a disease could simply drop out of the sky is a nightmare that few want to contemplate. The link with the Sahara became even stronger when the type of virus that caused the infection exactly matched the type O virus found in Africa's Sahel region around the fringes of the Sahara. Drought in that region had lasted 30 years and the land was badly degraded from overgrazing and infected with a host of diseases in a soaring population of cattle. 'There is absolutely no sewage system,' explained Dr Eugene Shinn, a US geological survey scientist. 'The contamination goes into rivers or the ground, that gets dried up and then blows away and spreads the germs.'

In fact, scientists are beginning to realise that the foot and mouth virus can make huge voyages across seas on strong winds, and could explain other seemingly inexplicable foot and mouth outbreaks. It was long thought that any virus carried up so high in the atmosphere would be killed off by the strong ultraviolet in the sunshine, but it is now clear that the dust shields the microbes from the ultraviolet sunlight.

Snow

4 June 1975

The weather of early June 1975 delivered an unbelievable shock – it snowed over much of Britain. It had been very chilly for much of May, and by June a wicked blast of cold Arctic air sent a biting frost across Scotland: on 2 June, Gleneagles, Perthshire sank to –3.3°C (26°F), more like the depths of winter than early summer. The *Daily Telegraph* reported 'Flaming June had arrived – freezing, and delighting only the skiers in the Cairngorms, where a blizzard replenished the slopes at Aviemore,' adding that one tourist, Mr Alan Friedland from Stafford, was actually delighted: 'It is the highlight of our holiday. I have taken pictures of my two young sons enjoying a snowball fight in June. This surely must be unique.'

The cold air swept down into England and snow fell as far south as East Anglia and London, with sleet reaching Portsmouth. Although the snow quickly melted in the south, it settled on the ground further north. Play was abandoned at the county cricket match between Derbyshire and Lancashire in Buxton, where snow reached an inch (2.5cm) deep. But it did not help Derbyshire because after the snow thawed they suffered a crushing defeat. Snow also delayed play between Essex and Kent at Colchester. John Arlott reported in *The Guardian* that snow also fell at a cricket match at Lords, London, although a complaint was made to the Press Council that he had exaggerated. *The Guardian* refused to print a correction or letter that, as the *Daily Telegraph* commented on, 'pointing out the absurdity of the report.' But meteorologist Professor

Gordon Manley stated, 'it was the first occasion in reliable records that snowflakes had been seen over London as late as the beginning of June'. The Press Council did not uphold the complaint.

Many other reports were given at other places of snow that day, as tourists shivered in their overcoats on all-but-deserted seaside promenades. The cold snap lasted a while, with snow lying on the ground for four days in parts of Scotland. But on 6 June the weather performed a breathtaking somersault and a heatwave sent temperatures soaring, followed by a gloriously hot summer across Britain.

Hampstead Flood

1975

The North London suburb of Hampstead is a chic place, famous for its stunning views over the rest of London from the high ground of Hampstead Heath. But it is also a focal point for some of the worst thunderstorms in the capital.

On 14 August 1975 Hampstead Heath was packed with visitors cooling off on a hot, humid and muggy day, when huge clouds towered high into the sky and by late afternoon the anvil shape over the top of those clouds signalled cumulonimbus in its mature and most dangerous stage. Cascades of rain crashed down with ear-splitting explosions of thunder and lightning, and in just over two hours, 6.7in (171mm) of rain fell and gushed down the hillsides.

The tiny Fleet, Westbourne, Tyburn and Brent streams grew into monstrous, raging torrents that ruptured their banks and sent waters cascading through gardens, buildings, basements and tunnels. Some houses were flooded over 6ft (1.8m) deep, and one man drowned in a basement flat. Hundreds of people were left homeless, and at Kilburn Park the Salvation Army set up a mobile canteen and council workers opened hostels for people flooded out.

The Underground flooded and one train was trapped for over two hours at nearby Swiss Cottage when floodwaters led to an electrical failure. The sewers could not cope with the deluge and manhole covers were blown open as fountains of water shot up into roads and flooded rush hour traffic. Yet the thunderstorm was so localised that just 2 miles (3km) away, the London Weather Centre recorded about 0.2in (5mm) rainfall from the same storm.

Hampstead had been caught between westerly and easterly winds colliding in an uprising of air. The heat on the ground and the high ground of Hampstead Hll gave an added push to the hot moist air, which cooled and condensed into cumulonimbus clouds.

This was no isolated incident – another disastrous flashflood struck Hampstead on 2 August 2002. But why Hampstead? It's no coincidence that such an exclusive area is so vulnerable to violent thunderstorms. The 400ft (120m) high hills that give Hampstead's expensive views also give the extra nudge to form cumulonimbus clouds.

Drought And Heatwave

1976

'Phew, What A Scorcher!' was the brilliant tabloid newspaper headline that summed up the summer of 1976. It captured the mood of a nation left wilting and bewildered by a Mediterranean-type heatwave. It still remains the hottest, driest and sunniest summer in the annals of British climatology, going back 349 years. And it also became the symbol of a new, hotter Britain.

Day after day, the skies remained cloudless and the sun beat down. Palls of smoke hung over the countryside where wildfires broke out. Sleeping at night was unbearable. And above all, water became a valuable commodity that could no longer be taken for granted.

The roots of the drought began the year before. A shortage of rainfall lasted through the spring, summer and autumn of 1975, and most critically over the winter. Without sufficient winter rains, the huge reserves of underground water that much of the country depends on were left unfilled. By the time the trees came into leaf the following spring of 1976 it was already too late. The ground was rock hard and the oven had been primed.

By March 1976, reservoirs in many southern areas were less than half full, and canals restricted use to avoid losing water losses through locks. April was bone dry, and although showers returned in May the rain could not filter through the parched ground.

The really intense heat set in on 23 June and set a new record for the longest and most intense heatwave in Britain.

Southern England had 15 consecutive days of 32°C (90°F) or more, and no heatwave before or since has come anywhere near this.

For nine weeks, the weather remained remorselessly dry, sunny and hot. As the ground baked hard, it helped stoke the heat like the lining of a furnace, re-radiating the Sun's heat soaked up by the ground. The so-called Great Drought saw reservoirs, lakes and rivers dry up and turn into giant cracked mosaics of mud.

The tinderbox conditions set off heath and forest fires, and many people were surprised to see snakes escaping from the infernos. When forest fires raged across the New Forest, firemen fighting the flames ran out of water. Lawns turned brown, crops shrivelled, and elm trees became so stressed from drought they succumbed to Dutch elm disease, a fungal infection carried by beetles that left the dead and dying trees dotted over the landscape like gallows.

Air-conditioning was still something of a novelty for many workplaces, and indoor conditions became intolerable. Passengers suffered from heat exhaustion on trains, buses and especially the Underground, and dozens of people collapsed at the Wimbledon tennis championships, which took place at the height of the hot weather. So freakish was the lack of rain that a crowd at Lord's cricket ground cheered when in mid-June a few drops of rain fell and stopped play.

Unusually for Britain this was a national heatwave. In Scotland in early July snowploughs were employed to spray sand on the bubbling tar of the A9 Perth–Inverness Road, while at Edinburgh Zoo the penguins had to be sprayed regularly with iced water to help them cope with the heat.

The peak of the drought coincided with the holiday season and as tourists poured in to the Southwest much of the region could not cope with the increased demand for water. As the drought gripped tighter, water supplies were cut off to homes in parts of East Anglia, Yorkshire, South Wales and Devon, and people were forced to use emergency standpipes in the street, almost a return to olden days of using a village pump.

Evoking the spirit of wartime economies, it became patriotic to save water – cars were left dirty, washing-up water was poured down the toilet instead of flushing and the golden rule for bathtime was to use 5 inches of water at most and then pour it onto the garden. The popular slogan that summer was 'Save Water, Bath With A Friend'. Hosepipe bans were enforced by patrol vans, and vigilantes enforced the rules in Surrey where housewives discovered a nearby golf club using water sprinklers and harassed the groundsmen until they turned their sprinklers off. British Rail stopped washing their trains, but people wondered if anyone would notice the difference.

Farmers watched in despair as more than £500m of crops wilted and it was a struggle to feed livestock. Wildlife was under threat as animals used to wet and mild British summers were left without food, water or shelter. But plagues of aphids crawled everywhere, quickly eaten up by swarms of ladybirds, which, when they ran out of aphids, started to bite people.

Dowsers searching for buried water supplies became extremely popular. In desperation, advice was sought from the US on artificial rainmaking by seeding clouds with silver iodide to induce rainfall, and Aboriginal rainmakers were brought in from Australia. Neither tactic worked. There were even thoughts of importing water in tankers from Norway.

As the ground dried out, thousands of householders were horrified to discover huge cracks in the walls of their homes, and a new word became part of the familiar vocabulary that year – subsidence. Homes built on clay foundations were particularly vulnerable; as the clay dried out it shrunk and the buildings cracked. Claims poured into insurance companies with costs estimated at £60 million.

One strange and unexpected result of the hot summer was a sharp decline in the birth rate in 1977. And significantly more boys than girls were conceived, which scientists believe may have been an effect of the heat on sperm, because the female X-chromosome is more sensitive to heat.

Two months later, and with no end to the crisis in sight, an emergency Drought Act was passed on 6 August. The Prime Minister, James Callaghan, appointed a drought minister, Denis Howell, whose homely advice was revealed as a brick in the lavatory cistern to save flushing away too much water.

However, the saviour of the water crisis turned out to be the August Bank Holiday weekend, when the heavens opened and the rains crashed down. But that did not end the drought for everyone, with water restrictions in Wales lasting for weeks afterwards.

It was quite possibly the country's greatest drought for 1,000 years and one of the greatest natural disasters in modern times. Those desert-like conditions were caused by a long run of high pressure systems that refused to budge, what is known as 'blocking highs' because they block out any other weather system.

The heat and drought stunned the nation. It was a shock to realise that Britain could experience Mediterranean weather and run short of water. London is highly vulnerable to drought, already one of the driest capital cities in the world, with a rainfall similar to Barcelona and available water per head of population equivalent to Israel. Two or three years of long drought revealed how easily the Southeast could run out of water.

Perhaps the lasting legacy of 1976 was a wake-up call to take climate change seriously. Until then scientists had been making noises about something going wrong with the weather, but it seemed so academic. After all, what was there really to worry about, when the first half of the 1970s had been a grim procession of cold winters and disappointing summers.

Crop Circles

Crop circles first caught the public imagination in 1980 when there were reports of cereal crops flattened into perfect circles in southern England. They were linked to strange noises, weird lights and were seen by some as evidence of visiting UFOs that had left their marks on fields of wheat. Questions were asked in Parliament, crop circle watches were set up with video cameras and radar to find one actually being formed, and the circles became an international sensation.

Then in September 1991 two pranksters confessed to making dozens of crop circles with nothing more than string and wooden planks. The credibility of corn circles was flattened.

But could some of the circles actually be a genuine natural phenomenon? Old woodcuts show circles of crops flattened by what was called 'devil's circles'. One of the earliest known illustrations was from 1678 in Hertfordshire inscribed as: 'The Mowing-Devil: Or, Strange News Out Of Hartford-Shire.' This described a field of oats reported to have been scythed down by the devil during the night. 'And so it fell out, that very Night, the Crop of Oats hew'd as if it had been all of a flame: but next Morning appear'd so neatly mow'd by the Devil or some Infernal Spirit, that no Mortal Man was able to do the like. '

In 1880, the science journal *Nature* carried a report from John Rand Capron, a well known meteorologist, who investigated some circles in a field near Guildford in Surrey. 'The storms about this part of Surrey have been lately local and violent, and the effects produced in some instances curious,' he wrote. 'Visiting a neighbour's farm on Wednesday evening (21st), we found a field of standing wheat considerably knocked about, not as an entirety, but in patches forming, as viewed from a distance, circular spots.'

Some idea of how a thunderstorm could produce such circles in fields of grain came to light in July 2002, when an eyewitness in Grand River, Prince Edward Island, described how he saw several whirling funnel-shaped clouds during a thunderstorm creating depressions in a neighbour's wheat field. Derrick Blacquiere discovered that the vortices left behind flattened circles up to 6m (20ft) across in the wheat. As meteorologist Mike Campbell at Halifax, Nova Scotia, explained: 'What we believe is a lot of air had rushed out of the thunderstorm and when it hit the ground it swirled around. It just whipped up the crops and flattened them into circles.'

So perhaps, in the words of 'The X-Files', the truth is out there, somewhere...

Great Balls of Fire

One of the great mysteries in weather is a rare phenomenon of thunderstorms known as ball lightning.

On 7 June 1996 a ball of light flew into a printing works in Tewkesbury, Gloucestershire. Workers at the factory watched in astonishment as the blue and white light bounced along the ceiling, passed through netting and whizzed round girders, hitting machinery and sending sparks flying. Finally the light hit a window and exploded with an orange flash and a tremendous bang, knocking out the telephone switchboard. 'The whole place was lit up,' said Simon Pocock, one of the staff. 'The sparks were unbelievable – it was like a horror movie.' Three people received electric shocks, including one woman who was hit in the shoulder. One worker said they had seen the ball of light seem to roll off a jet plane flying overhead that had been struck by lightning.

Another bizzare and terrifying incident struck on 25 April 2004, in Brecon, South Wales. During an intense thunderstorm, a huge white globe was seen falling out of ink-black clouds over the town. It struck a house, blowing up the chimney and showering bricks over 30m (100ft) away, while the ball itself disintegrated in a shower of small lightning bolts that shot over buildings across a quarter of a mile. Moments later a crack of thunder sounding like a bomb going off shook the ground and sent people diving for cover in fear of their lives. Soon after that, a streak of blue-coloured light was seen shooting through an office with a loud crackling noise, before it hit a pipe, bounced backwards and then disappeared. Across the town, shopping tills, computers and televisions were badly damaged. It was amazing that no one was injured.

These incidents may sound like something out of 'The X-Files', but were possibly exceptionally violent cases of ball lightning, a rare and mysterious type of lightning. No convincing explanation has been given for the phenomenon, but it seems to be formed in intensely electrified air, usually during severe thunderstorms. For years mainstream scientists dismissed it, but now there are so many accounts of ball lightning that it has become accepted as a real phenomenon, although incredibly difficult to study. Usually the ball of light is fairly benign, sometimes gliding slowly through the air, able to pass through closed windows and often drawn to electrical equipment, such as televisions or light switches.

Ball lightning has even been seen passing down the aisles of aircraft and could also explain a phenomenon seen during the Second World War. Many pilots reported balls of light chasing or clinging to their aircraft, and fearing they were being attacked by an enemy weapon, they tried taking evasive action, but without success. The phenomenon was called 'foo fighters', but caused no harm and has remained a mystery ever since.

All in a Spin

In June 2002, the Queen and Duke of Edinburgh were at Royal Ascot races when they, and all the other racegoers, watched in horror as an amazing whirlwind spun and ripped through the racecourse. A gazebo, chairs, picnics and ladies' hats were sent flying some 30m (100ft) high into the air as people scattered out the way. The Met Office estimated the vortex was gusting to storm force 6 for a short time.

Whirlwinds such as this are called dust devils because they often kick up clouds of spinning dust from the ground, and are quite different from tornadoes, which are spawned from thunderclouds. Warm air rising up off hot ground can start twisting like a corkscrew as it penetrates cooler air above. As the winds squeeze into a tighter vortex they spin faster, and can last from a few minutes to several hours.

The country town of Devizes in Wiltshire received a baffling shower in 1982, when masses of hay fell from a clear blue summer's day. 'Late afternoon a great cloud of hay drifted high over ahead, then dropped onto the centre of town,' reported the *Bath and West Evening Chronicle*. The airborne haystack was created by a dust devil sweeping up straw from a field and scattering it far and wide.

Sometimes dust devils can inflict serious damage. Possibly the most expensive whirlwind was at an antiques fair, the Detling International Antiques and Collectors Fair in Kent in July 1999. Around 3,000 people watched in horror as stalls were sent flying. 'The tornado came out of nowhere,' said Sue Ede, the director of the fair. 'We were on the showground and a spiral of air about 50 to 60ft high suddenly whizzed straight through the middle of the exhibitors. It lasted for 30 to 40 seconds, but it seemed like two hours.' No one was injured, but many of the antiques were damaged. 'It is not the kind of thing you expect on a beautiful sunny afternoon in Kent,' she said. 'The first thing I thought I saw was a parachutist in the sky, but then I realised it was part of a marquee flying off into the air.' The whirlwind came at them so quickly that the crowd had no time to panic or try to run away, she said. 'People reacted in a typically British way afterwards and were worried about the person next to them, before starting to pick up the pieces of broken china and antiques around them.'

Heat Record

2003

The whole of Britain rarely has the same weather, but in August 2003 the entire country was roasted in an astonishing heatwave. Each day the thermometer inched upwards, until on 10 August Britain's highest temperature record was set with 38.5°C (101.3°F), at a weather station sited at the National Fruit Collection at Brogdale near Faversham in Kent. This was the first time 100°F officially was broken in the UK. 'This is a heat wave like no other,' *The Times* exclaimed, and it was true. Only the previous day, Scotland broke its own national record with 32.9°C (91.2°F) at Greycrook on the Borders.

To put it into perspective, temperatures reaching 36°C (97°F) are so rare in Britain they have happened only twice before in the past century: in August 1990 and August 1911.

This new temperature record was a symptom of an unmistakable trend – summers in Britain are growing hotter. The hottest months for the 20th century tell the story: the 1990s had six of the hottest months while the 1980s and 1970s had three each, and the 1960s had none.

So the summer of 2003 was truly groundbreaking, not just here in the UK but across Europe, where it was reckoned to be the hottest summer in at least 500 years.

August 2003 highlighted the lethal dangers of heat. About 2,000 more people than normal died during the heatwave in England and Wales, and 27,000 across continental Europe. That experience led the British Government to issue a heatwave plan with four levels of alert. Measures include staying out of the sun, avoiding the outdoors from 11am to 3pm, wearing loose clothes and a hat, taking cool showers, and consuming cold food and water.

The heat was particularly intolerable in London. The London Eye was closed when the heat inside the pods became unbearable. The heat and sunshine sent air pollution soaring, and doctors warned people not to go out jogging. Air conditioning failures forced the closure of factories and offices. Satellite heat pictures revealed that London often now has the climate of Nantes in the Loire Valley thanks partly to the heat given off by the capital's concrete, bricks, tarmac, and other urban environment turning the climate hotter, especially at night when the heat soaked during the day is given off just like a heat storage radiator.

Impacts of the Heat

The impact of the Mediterranean-type heatwave of 2003 rang alarm bells, and for months afterwards its impacts were closely scrutinised for clues as to what the future might bring as Britain becomes engulfed by climate change.

It highlighted a looming water shortage crisis for London and the Southeast. London has an average annual rainfall less than Madrid or Istanbul, and as its population rises, the water resources are under increasing strain. To make matters worse, Britain has inherited a decrepit Victorian water supply system that leaks like a sieve, and even in the great drought of the 1890s was leading to cuts in water supplies (see 'Drought 1893'). Now some 3.5 billion litres of water are lost each day in England and Wales, more than a fifth of all the country's supply. London has the worst leakage record of all, and in 2005 the London Assembly warned that the capital could run into shortages of water in ten years.

The building industry is concerned that most British houses are not designed for heatwaves and many homes will turn into ovens as indoor temperatures soar above 25°C (77°F), a key threshold when people feel very uncomfortable. New houses need to be designed to be much cooler, using better natural ventilation, more shading, and denser walls and floors, which can absorb heat more easily. The fear is that more people will turn to air conditioning for a quick fix, with the result that power demands during the summer could rocket and outstrip supply, leading to power cuts. Already, power consumption from air-conditioning has quadrupled over the last 20 years.

Roads could melt. Rather like the way toffee melts in hot weather, roads will disintegrate as the bitumen-based binder used to hold the surface aggregate together starts to melt. This increases the risk of skidding, and will also lead to ruts, potholes and eventually the top surface of the road could peel off under the onslaught of traffic. Concrete-surfaced roads will be in danger of buckling as the slabs of concrete expand in the heat.

Bridges can seize up as their expansion joints, used for taking up the movements caused by heating, no longer work. This means much more expensive maintenance, a big concern for local authorities and the rail industry.

Train speeds will be severely reduced as rails buckle. Current safety standards demand that between 36°C and 40°C (97°F and 104°F) trains have to slow down, depending on the normal speed limits of the track. Over 40°C (104°F), the speed restrictions are more drastic, causing further delays.

Animal and arable farming could be badly hit. Chickens are particularly sensitive to heat, and in August 2003 thousands of poultry died from heat exhaustion. Pigs also need better ventilation or air conditioning in intense heatwaves. The hot weather hits crops such as strawberries, which can ripen too fast, peas become too tough and potatoes can be of poor quality.

Ozone air pollution is forecast to increase dramatically, especially in the countryside. Ozone is made by sunshine cooking up traffic and industrial pollution, as well as natural volatile chemicals given off by vegetation. This will exacerbate respiratory problems such as asthma and bronchitis. Ozone pollution can actually be worse in the countryside than cities, causing damage to crops.

Heat also affects human behaviour. Research shows that rising temperatures lead to more road rage incidents, but it also causes more lethargy and absenteeism from work. Relate, the marriage guidance service, says calls to its helplines increase by up to 30 per cent in hot weather. Psychologists point to heat as a big factor in the seasonal rise in violence, including an increase in suicides. There are many other factors involved in this violence apart from heat, though, such as more people outdoors and more alcohol consumed. However, violent crime and riots increase as temperatures rise, and the majority of riots in the USA occur when the temperature increases to between 27°C and 32°C (81°F–90°F). No surprise, then, that the murder rate also peaks in the summer months.

Quirky Sides of Heatwave
2003

White wine, cider and lager sales all increased, as expected, but Sainsbury's also saw sales of sherry shoot up 65 per cent over the hot summer, apparently washed down as a cool drink with ice cubes.

As expected, sales of fizzy drinks soared, but only up to around 23°C (73°F), at which point bottled water sales rose instead. More chicken sandwiches than cold beef ones were eaten, more tea than coffee, and at 20°C (68°F) sales of hot food at motorway service stations nosedived. Safeways reported selling eight times more burgers, seven times more iceberg lettuces and four times more sun cream. By far the biggest impact of the hot weather was on sales of leg wax, which increased some 14-fold, closely followed by hair removal cream.

A worrying feature of the hot summer was a huge rise in sales of air conditioning units and fans, placing big demands on power supplies at a time when many power stations were taken out of service for maintenance in the summer.

Spending more time outdoors also explained why one department store reported a 73 per cent surge in the sale of trampolines.

People had more tattoos and piercings as more skin was bared and wanted it decorated.

Umbrella suppliers were hit hard, but the lack of rainfall led to increased sales of water butts.

Despite the lack of rains, wheat yields actually went up and with good quality grain farmers' profits soared by £30m. Barley grew better, too. Other vegetables also benefited, but sugar beets and field beans showed small losses. The negative impact on livestock was more severe. The weight of home-fed sheep and pigs dropped, with financial losses, and milk yields fell some 50 litres per cow below average, costing dairy farmers about £10m.

English vineyards enjoyed a vintage year with high yields and 'fantastic quality'. The mild spring meant that the vines were left undamaged by frost and the dry weather that followed provided a fantastic summer ripening period. Harvests came in a fortnight earlier than expected.

Concorde flights were slowed down because the extreme temperatures experienced in the UK and across Europe made the air less dense and less able to support a supersonic aircraft.

The amount of fuel Concorde carried was cut to get the plane airborne at Heathrow, forcing the flight to refuel in Newfoundland.

Britain's transport system suffered, particularly the railways. Widespread speed restrictions were imposed on the railways because rails buckled, which became a real problem when rail temperatures reached 36°C. Fires on the side of the tracks jumped 42 per cent.

Britain's road network also suffered in the searing heat. Sections of the M25 melted and the total costs of repairs across the country's road network were estimated at £40m. Temperatures on the London Underground passed 41°C and passenger numbers dropped, reducing revenue by £500,000.

The numbers of wasps and ants soared.

Birmingham Tornado

2005

Tornadoes hardly figure as one of the big natural threats that Britain faces. But the country is hit by an aervage of 60 tornadoes each year. Of those, an unusually high number hit the West Midlands, where Birmingham could lay claim to being the tornado capital of Britain, hit by an astonishing 32 tornadoes in just over 70 years. Although most of these are fairly weak affairs, some can reach dramatic proportions.

One of the worst incidents struck on 28 July 2005. In a scene out of the 'Twister' film, shop fronts crashed in, roofs were ripped off, buildings collapsed, over 1,000 trees felled and 19 people injured. Damage was estimated at £39 million. This was a disaster more familiar in Tornado in the Midwest USA, than the Midlands.

The tornado's wind speeds were estimated at over 140mph (225km/h), far more violent than the storm that hit Britain in October 1987. People watched in horror as a huge black funnel cloud bore down on the city.

'There was a little bit of thunder and then some lightning. The winds started to pick up and then tiles began to fly off roofs and smash into windows. Huge trees have been uprooted and complete roofs have been ripped off houses. Smashed cars, uprooted trees and lampposts are lying all over Sparkbrook. My sister actually saw the cylinder of the tornado going through the sky,' described Mrs Reefat Batul of Sparkbrook.

Jane Trobridge working in Small Heath: 'Suddenly everything started blowing upwards. I tried to get out to get a better look but I couldn't actually force the door open. I could see loads of flying debris circling, and then the roof seemed to lift off the petrol station and swing almost completely upright and then fall back down. Behind it there were trees and bricks flying about. It was madness; cars being lifted up; flag poles snapped off at their bases.'

The devastation looked like a bomb had exploded: houses ripped apart and streets littered with glass, bricks, furniture and smashed cars. A path of destruction over 7 miles (11km) long and up to some 550yd (500m) wide was carved through Kings Heath, Moseley, Balsall Heath and Sparkbrook. It was amazing no one was killed. Some homes were still uninhabitable a year later and the regeneration of Balsall Heath, one of the most deprived wards in England, was severely set back.

Only three months after the July 2005 disaster, another tornado struck on 12 October within a mile of the big one, although it was far weaker and caused little damage.

Studies by the Tornado and Storm Research Organisation (Torro) revealed that Birmingham has been hit by even more violent tornadoes in the past, and could well be struck again. In 1931, a tornado with winds estimated at over 160mph (260km/h) tore through a remarkably similar area of Birmingham, almost flattening some buildings. One woman was killed and several badly injured, although casualties would have been much worse had torrential rain beforehand not sent people running for shelter.

So, is Birmingham and the Midlands the Tornado Alley of Britain? A study by Peter Kirk of Torro shows that over 25 years the West Midlands has the third highest number of reported tornadoes of anywhere in the UK. 'But this could be because so many people live in the area, so they are more likely to report a tornado,' he added.

The geography of Birmingham may also play a part. The sort of ferocious summer thunderstorms that can spawn powerful tornados are more likely to develop inland, particularly in central England, away from cooling seabreezes. The thundery weather tends to sweep up from the Bristol Channel or South Wales, and may get an extra lift from hills to the southwest of the Birmingham conurbation and set them rotating.

In fact, a further study discovered that the July 2005 tornado was spawned by a type of thunderstorm called a supercell, more often seen in Tornado Alley, in the Midwest of the

USA. These storms are much more powerful and last longer than an ordinary thunderstorm, driven on by high-level winds. The day of the tornado in July 2005, hot and humid air from the tropics collided with colder, drier air from the north, and created big thunderstorms. As winds at different levels in the atmosphere rushed in from different directions they set the storm cloud rotating and, deep inside the cloud, a narrow core of air spun even faster until eventually it descended from the thundercloud as a tornado.

Much remains to be learnt about how tornadoes form – why, for instance, only a few supercells spawn a tornado, and what part the geography of the landscape plays. This is why it was virtually impossible to give the Birmingham residents any specific tornado warning. 'Tornadoes are part of the British climate, and some are likely to be as strong as the Birmingham tornado,' explained Paul Knightley of Torro. 'What was so unusual is that it hit a large populated area where it caused huge damage.'

In fact, it was only a year later that another terrifying twister blitzed London. On 7 December 2006, a tornado tore through Kensal Rise, northwest London, with winds of around 100mph (160km/h), leaving about a hundred properties damaged. Rooftops were ripped off and cars were badly damaged, one man suffered a serious head injury and five people were treated at the scene for minor injuries and shock.

Amazingly, fatalities from tornadoes in Britain are extremely rare, though. The worst incident was six people killed at Edwardsville, a mining town in Glamorgan, on 27 October 1913. 'Men were lifted from their feet and dashed to the ground' according to one press report. The headmaster of the local boys school wrote a report of his experience. 'We heard a noise resembling the hissing of an express locomotive. The sound grew rapidly in volume, at last resembling the rushing speed of many road lorries racing along. The oppressiveness that had been previously noticed increased, and the heat and air-pressure were pronounced during the rushing noise.... the panes of our window were broken by stones, tiles, slates, dried cement and splintered timber.' Over a hundred people injured, houses and a chapel were wrecked and roofs were scattered for several miles.

Wimbledon Tennis

Wimbledon tennis is often cursed by rain in June. It is extremely rare for the whole Wimbledon fortnight to escape wet weather, and a whole day's play has been washed out 30 times in the tournament's history. Sixteen times since the Championship moved to its present ground, play has been extended into a third week to clear a backlog of matches rained off. Or to put it another way, a completely dry Wimbledon happens on average only once every 20 years.

Part of the reason for this wet record is thanks to Britain's monsoon season. Usually our wet westerly winds sweep bucketloads of rain off the Atlantic and these winds tend to die off in the springtime, often leaving the country fairly dry. But by mid-June, just in time for Wimbledon, the wet westerly winds often return and can drag on until the end of the month. The weather generally improves by early July as high pressure exerts its influence.

History Of Wet Wimbledon

1922. The championships had an ominous start when they moved to their current grounds at Church Road, Wimbledon, and were plagued by rain every day. At that time, only the Centre Court was protected by tarpaulins, and so the other courts turned into quagmires. Heavy downpours during the second week resulted in a huge backlog of matches, and the tournament was forced to squelch into a third week, with the last ties not completed until Wednesday 12 July. This ranks as the most disrupted championship in Wimbledon's history.

1927. Heavy rain on most days forced play to be extended from 12 to 14 days.

1958. One of wettest first weeks ever, but the tournament managed to finish on time.

1968. 1 July so hot that a record 550 people need medical treatment for heat.

1972. The second Saturday was washed out with four finals played on the last Sunday for the first time.

1976. A tropical heatwave roasted London in its hottest fortnight on record, in what became a record-breaking summer drought. The grass brown, the courts turned to dust, spectators fainted in the heat. The average afternoon temperature for the 12 days of competition was 31°C, and the baking sun beat down unremittingly for an average of 12½ hours per day.

1982. The wettest Wimbledon, in terms of the sheer quantity of rain, about three and a half times the average amount for two weeks in midsummer. The single heaviest downpour crashed down with a staggering 3in (77mm) of rain in one day, almost double the average rainfall for June.

1985. Rain fell on every day in the first week, and especially on the first day when a thunderstorm rocked the grounds. 'Suddenly there was an enormous explosion. We were all wondering what the hell it was, it sounded just like a bomb,' recalled Chairman of the All England Lawn Tennis and Croquet Club, Buzzer Hadingham. Lightning had blasted a huge lump of masonry off the main building and landed only a couple of feet from a girl.

In the second week, the Met Office's newly installed rainfall radar revealed an approaching cloudburst. Forecaster Bill Giles phoned the tournament referee to warn him, and although the skies were cloudless over Wimbledon at the time, the groundsmen managed to cover the courts. Just 15 minutes later black cloud dumped a staggering deluge of some 1.5in (40mm) rain in just 20 minutes, swamping the spectator's and players' tunnels leading on to the Centre Court with 2ft (60cm) of water.

1986. Jimmy Connors dropped his racket after receiving an electric shock from intense electricity in the air just before a thunderstorm erupted.

1991. After four days of rain interrupted play, only 52 of 240 matches were completed. Matches were played on the middle Sunday, but the championship managed to finish on time.

1996. Rain poured down for three successive days, and when downpours of 'monsoon' proportions stopped play on Centre Court, Sir Cliff Richard entertained the sodden crowd with renditions of Summer Holiday with backing from Pam Shriver, Virginia Wade and Martina Navratilova.

1997. It was so chilly that on 27 June the mid-afternoon temperature sank to just 11°C (52°F). To add to the misery, two days were lost to rain in the first week, and the middle Sunday was played for only the second time. But the tournament ended on time.

1998. Another wet and cold WImbledon.

2001. Rain ruined the final Saturday and Sunday, with Tim Henman playing Goran Ivanisevic in semi-finals over three days. The final was played on the third Monday.

2004. For only the third time in its history, matches were held on the middle Sunday of the championship after rain hit the first week of play.

2006. The first day of play was almost a complete washout.

2007. Wettest Wimbledon for 25 years. Masses of rain in a record-breaking May–July rainfall sent the schedules into a tailspin, but desperate efforts were made to avoid playing beyond the usual two weeks.

Mirage English Channel

The atmosphere can play weird tricks with the light and create some fantastic sights of Alice in Wonderland proportions.

On Sunday, 8 August 2004 the Coastguard at Southampton received calls of a flashing light seen out at sea off the northern coast of the Isle of Wight. It seemed a ship was in distress, possibly on fire, but the light was flashing in such a regular pattern it made no sense as a distress call. The Coastguard called vessels in the area, but the only flashing light they reported was the lighthouse at Barfleur, near Cherbourg on the French coast.

To add to the mystery, other smaller lights began to shine as the sun set. David Butler, an auxillary coastguard at Ventnor, climbed up on to high ground and with binoculars could see clearly not only Barfleur lighthouse, but also the houses and harbour walls of Cherbourg, which he knew well.

> 'We sat and watched for the flashing light for 30 minutes, it was not moving, just a regular flash. We were a little bit perplexed and went on to higher ground, about 700ft above sea level and we could see the light clear as a bell. There was lots of shipping, but it wasn't shipping – it was houses. It looked like the harbour of Cherbourg and with binoculars, you could see the glow of town, outer buoys, harbour walls, buildings behind harbour walls, houses lit up, tall buildings, lighthouse. I know Cherbourg well, I most definitely know it well. I thought I'd gone completely bloody mad.'

Cherbourg is well over 60 miles (96km) from the Isle of Wight and is usually hidden under the horizon. But the amazing sight was revealed by a mirage, created by layers of air with different temperatures bending light over the horizon like a colossal glass prism.

Intense mirages may explain tales of lost islands in olden days, many of which appear on maps but were later found not to exist. One famous expedition in 1818 tried to find the fabled Northwest Passage around the Arctic north Canadian coast, and reported no passage existed because of a chain of mountains named the Crocker mountains. But a later expedition found no such mountains existed, they were a mirage.

An even weirder mirage was sent to *The Times* on 11 July 1860 by a reader on the coast at Edinburgh. 'Our attention was first drawn to the singular appearance of the Haddington coast [far away] beyond the visible horizon,' he wrote. 'A blue band, apparently of calm water, reflected everything twice its own height, and showed islands distinctly we had never before observed on the clearest days.' Into this strange spectacle appeared a ship steaming across the horizon. 'A ship of war with two decks, probably Her Majesty's ship Edinburgh... she was reflected double, as if two ships were sailing one above the other; and still further down this reflection became inverted, and she appeared to stand on her masthead.' The correspondent realised that the sight was a mirage, adding that 'the trees opposite were magnified into gigantic forms, and the houses distended into factory-chimneys, with windows all down them.'

It is probably not taxing the imagination too far to imagine that mirages could well explain some UFO sightings, although before the term flying saucers was invented, reports sometimes spoke of flying ships.

Summer Floods

2007

Trying to predict a summer months ahead is a very tricky job, and in 2007 it was unbelievably difficult. The spring was gloriously hot, and April was the hottest April on record, as well as exceptionally dry and sunny. Spring flowers came and went with indecent haste, and expectations were raised of a hot summer, with warnings of drought and water shortages on the way.

But the weather had lulled the nation into a false sense of security. Early May saw cloud building up from the Atlantic, the sun disappeared and the rains arrived. By late May bank holiday it felt like winter had returned: cold, wet and even sleet in places. It was one of the wettest Mays on record.

Summer showers can be very heavy but they tend to be short-lived and often local. The rains that summer were quite different. They were strung out across the country in long weather fronts that were slow-moving, giving plenty of time to drop their cargoes of rain. Glastonbury music festival in Somerset disappeared in a mudbath and Wimbledon tennis was a washout.

On 15 June parts of Yorkshire and the Midlands flooded and a couple of weeks later Sheffield was under water when the rains returned. Hundreds of people were stranded in flooded offices and shops in the city centre. The giant Meadowhall Shopping Centre was under 7ft (2.1m) of water in places, and one of those trapped described cars floating down the road outside. Rescue helicopters airlifted to safety those in most need. There were fears that the large Ulley dam outside Sheffield might burst when it showed alarming signs of cracking up, and the M1 was closed for a few days while engineers pumped out its water and reinforced the dam walls.

Hull was hit twice severely. One of the most deprived areas in England, the city also largely lies at or below sea level, and without drainage it fills up with rainwater like a bathtub. To combat flooding, a large drainage and pumping system jettisons the water into the River Humber. But the scheme was overwhelmed on 25 June when one-sixth of the average annual rainfall fell in a day, an event only expected once every 150 years. The drains became filled to capacity and although the pumps worked flat out they could not drain the rainwaters fast enough, so the floodwaters backed up the drains and filled large parts of the city with water and sewage. People found water seeping through their floorboards, as one resident described: 'It's as if there's a tap on. When you're squelching on the carpet you realise you have to move things upstairs.' The council's emergency plans for a flood were designed for the rivers overflowing, not for a huge rainfall event. More than 8,000 homes were flooded, 92 out of 99 schools were swamped and one man died trapped in a storm drain. Thousands of people were forced to live in temporary accommodation including caravans, and a year later many were still homeless.

July was even worse. The ground was now so saturated that any rain ran off the surface in flashfloods. A brutal weather front on 20 July dumped more than a month's rain in one day in many places, reaching 5.8in (147mm) in Sudeley, Gloucestershire. The rains seemed to come down in waterfalls that ricocheted off the ground in a cloud of mist.

Parts of London were under water, some areas of the Underground were flooded and dozens of flights cancelled at Heathrow Airport as runways disappeared under water. Roads quickly became congested in a massive traffic jam across southern Britain, just as the school holidays began in many areas.

Over the following days, the Severn, Avon and Thames rose to bursting point and swamped a large swathe of central and western England, exceeding even the disastrous floods of March 1947.

Gloucestershire was worst hit. Half a million people were close to losing their power supply when an electricity sub-station came within inches of being swamped, only saved by hundreds of troops and fire service personnel working round the clock to build a flood barrier and pump out floodwaters. But a water treatment plant was overwhelmed and more than 350,000 people left without fresh water for more than two weeks. Hundreds of small water-tankers were brought in to supply water, but when these could not cope with demand, empty beer tankers were used to ferry in more water. The Red Cross distributed hygiene kits.

The entire summer floods left 13 people dead, destroyed 48,000 homes and 7,000 businesses in the South West, the Midlands, Yorkshire and Humberside. Power and water supplies were lost, railway lines washed out, eight motorways and many other roads were closed and large parts of five counties and four cities were brought to a standstill. Whole towns resembled islands surrounded by water. It was the biggest loss of critical infrastructure since World War II.

The insurance industry was faced with total damages estimated at £3 billion, with about 165,000 insurance claims, the biggest single claims event in British history. As a result, insurers raised their homes premiums to recoup their costs. And the cost to the economy of dislocation from the floods was probably a further £3 billion.

It was the wettest May–July period in records going back to 1766. Yet the summer's rains seemed to come out of the blue, after 30 years of increasingly hot and dry summers. The immediate blame was pinned on the jet stream, a river of wind a few miles high that snakes around the world. The jet stream steers depressions across the Atlantic and in the summer usually drifts north of the UK, taking its wet weather with it. As the jet stream departs, its place is often taken by a block of high pressure from the sub-tropical Azores, which usually pushes north and bathes the UK in sunny, dry weather. But in 2007 the jet stream swung unusually far south. It was also sluggish and buckled into an enormous loop that parked bands of rain over the UK with little to blow them away. On the other side of the jet stream, southern and eastern Europe scorched in blistering heat with huge forest fires in Greece. But why the jet stream took a southerly track remains a mystery and underlines our lack of knowledge about so much of the world's weather.

'This summer was unprecedented,' explained Dr Terry Marsh at the Centre for Ecology and Hydrology, Wallingford. 'The wetness of the soils and the river flows in the lowlands of England are more typical of winter than summer, I've never seen anything like that before.'

The Great Flood of 1947 was the previous worst inland flood in Britain, which submerged 700,000 acres (2,835sq km) of land and caused an estimated £4bn worth of damage, in today's money. But that was triggered by a combination of rapidly melting snow, a frozen ground as well as heavy rains.

'There were no close modern parallels to the scale of the summer flooding,' concluded a report into the floods by the Centre for Ecology and Hydrology. Weather historians had to go back to soaking wet summers such as 1912, the wettest and dullest summer on record. But the closest parallels were probably the summers of the 1800s, which were often wetter than the winter, the opposite of rainfall patterns over the past century. The summers of the mid-1840s were incredibly wet, which triggered the horrendous potato blight epidemic of 1845 onwards that wiped out the staple crop of Ireland and led to mass starvation. In the 1870s, the country suffered another run of exceptionally wet summers, leading to a run of crop failures that tipped British agriculture into a recession that lasted decades.

Of course, someone or something has to be blamed for lousy weather. Two centuries ago, wet summers were blamed on the cannon fire in the Napoleonic Wars, the rains during the First World War were the fault of artillery going off on the Western Front, and another run of poor summers in the 1950s was even thought to be caused by nuclear bomb tests. And now it's global warming, although there's no evidence that the rains were anything to do with climate change.

However, an interesting feature of summer 2007 was how the temperatures held up surprisingly well despite the clouds, rains and floods, whereas wet summers of long ago were invariably cold. Perhaps this underlines how the underlying pattern of rising temperatures continued even in such a dismal summer. And the severity of the rains were consistent with predictions of more violent bursts of rainfall in climate change.

Effects of the Floods

Farms were badly hit. Thousands of tonnes of crops were left rotting in the saturated ground: some 40 per cent of the UK's pea crop was lost and 20 per cent of potatoes damaged by floods and potato blight, with farmers unable to spray the crops because the ground was too wet. Supplies of rapeseed, broccoli, courgette, cauliflower, sweetcorn and parsnips were all affected by the torrential downpours and flooding. Cows spent much of the summer indoors, forcing farmers to use scarce silage stocks, sending feed costs higher as grain prices increased.

Most retailers had a wretched time. The sunshine in April lulled clothing stores into ordering summer clothes, which went largely unsold. Stuart Rose, the chief executive of Marks & Spencer, came to regret his comment made the previous year, that retailers should never blame the weather for poor results – by doing just that when he reported sales in July 2007. 'There's weather, and there's weather,' he said, adding that when you see pictures of people wading through flood water 'you've clearly had extreme weather'. However, sales of vests, macs and women's weatherproofs soared by 150 per cent.

Obvious losers from the floods included gardening centres, air conditioning installers, barbecues and fridges. But there was a big rise in sales of DVDs, tumble dryers, mops and buckets. Fake tan also proved popular.

The dire weather cast an interesting light on the behaviour of the British. A survey of 1,333 people by a supermarket chain found 18 per cent were eating more chocolate, 14 per cent more cakes and 9 per cent more pasta and bread as a result of the bad weather. By contrast more than a quarter stopped eating salads altogether. Nearly two thirds blamed the grey days for increased feelings of lethargy and irritability, while one quarter abandoned exercise regimes. Almost one third believed lack of sunshine lowered their productivity at work. Forty-three per cent stayed at home rather than socialise in order to avoid the rain.

As expected, sales of ice cream, soft drinks, salads, sandwiches, barbecue food, raspberries, strawberries and other soft fruit, beer and cider all slumped. Britons abandoned healthy salads and fruit and instead sought solace in stodgy, comfort foods, such as ready-made meals, tinned soup and mince – sales of meatballs shot up 20 per cent. 'People have been buying comfort food and what they usually buy in October, not what you would usually expect them to buy in August,' said one retail analyst.

The damp weather also saw an upsurge in cough, cold and flu medicines sold, although there was a 40 per cent drop in allergies, such as hay fever. Cinemas and nightclubs reported brisk business and Rank's bingo halls and gambling website also drew lots of punters in. Small surprise that late holiday bookings abroad boomed, whilst tourism in Britain took a terrible hit – some attractions had only a quarter of the normal number of visitors, and a number of businesses were forced to close down and many others fought for survival. But tourists arrived in force when the dried-up source of the Thames, near Cirencester, briefly spurted back into life. The Thames Head Inn did a roaring trade.

Wildlife was affected in some unexpected ways. Demand for Rentokil, the UK's largest pest control company, rose by more than 25 per cent when the floods drove rats above ground, with many invading houses.

There was a severe outbreak of ticks and fleas in the countryside, putting people and animals at risk of potentially fatal blood diseases, with Hampshire reporting 220 incidents of Lyme disease. There was a boom in all sorts of creepy crawlies, including woodlice. The wet and mild temperatures triggered an explosion of mosquitoes, and the NHS Direct phone helpline reported abnormally high numbers of people reporting insect bites during August. A particularly worrying development was the reported sighting in Somerset of the Asian tiger mosquito, which can carry the potentially fatal West Nile and Dengue disease viruses. September brought no relief as the warm and wet weather led to a boom in wasp numbers and Rentokil reported a 125 per cent national rise on the previous year in call-outs to deal with wasp nests, some of the monsters up to four times the size of a normal nest.

Gardens and farms faced a plague of slugs and snails. Slugs reached record numbers with almost 15 billion estimated across the nation, up by well over 50 per cent. Conditions were so good that many species were able to fit in one or two extra breeding cycles.

But for many other wildlife it was a grim summer. From water voles to swallowtails, partridges to bumblebees, many creatures suffered catastrophic losses, especially of their young. For example, the cold and wet hit the grey partridge so hard it turned from being a common and familiar bird to a decline of nearly 90 per cent in Britain. Blue tits and great tits suffered their worst breeding season on records. Reed warbler, whitethroat, willow warbler, treecreeper and willow tit had the lowest number of surviving chicks since bird numbers were recorded 25 years previously. The young died because they did not have enough feathers to protect them from the heavy downpours and there was also a shortage of food for birds that rely on eating caterpillars, because the rains washed the caterpillars off trees. Water Voles were drowned in swollen rivers in their burrows. And the wet weather hit bats so hard that many mothers were forced to abandon their young in a desperate search for food.

But for plants it was a wonderful season. After years ravaged by heat and drought, the summer of 2007 marked a revival for trees, which later put on a magnificent display of autumn leaf colours. Grasses, flowers, mosses – almost all thrived in the wet climate, apart from those flooded out. There was also a big increase in fungi, which appeared earlier than usual.

The storms had far-reaching effects far away from the floods. The huge volumes of rainwater washed pollution into the sea, resulting in a steep increase in the number of polluted beaches.

Outlook

After years of concern about droughts and heatwaves in summer, the pendulum swung the opposite way, and instead of demands for more reservoirs and water supplies there were cries for better flood protection. The disastrous floods particularly raised concerns about building new homes on floodplains, and the Government's commitment to building three million new homes by 2020, many of them in areas liable to flooding.

A report into the floods highlighted the need to flood-proof emergency services, electricity and gas sub-stations, water and sewerage treatment centres must be prioritised. Buildings earmarked for evacuation and rest centres in such incidents also needed to be resilient to flooding. One suggestion was that planning permission should be required before front gardens could be concreted over to provide off-street parking or for new garden sheds and patios. The run-off from these hard-surfaces was seen as a significant problem in the flooding, particularly in Hull and Sheffield.

Flood awareness should also be part of the home-buying process and people should know in advance if the prospective house were at risk.

Another suggestion was that every home in flood-prone areas should also keep its own emergency flood kit. A sealed plastic box should be stored as high as possible with essential documents, emergency contact numbers and items of sentimental value, and equipment such as torches, radios, blankets and first aid kits.

AUTUMN

Julius Ceasar Invasion

54BC

The Romans had a tough time trying to conquer England. Julius Caesar launched their first invasion in 55BC, and although it was not very glorious, he did leave us the first written weather report of England.

In later summer, a Roman fleet of about 80 ships carrying 10,000 infantry set sail at night from near Boulogne in what was then called Gaul. Conditions were fair for the Channel crossing, and next morning the Romans could see the cliffs of Dover, although English warriors were waiting for them on the clifftops. So the invasion force sailed further up the coast looking for a safe landing place, and found flat beaches near Deal, where the Romans established a camp. But four days later, 'such storms ensued that the task [of fighting the English] was of necessity interrupted and the continual rain made it impossible to maintain the troops in tents,' Caesar wrote in *de Bello Gallico*. In high tide and a storm surge the Roman ships were battered and many wrecked. Those vessels that could be repaired had to be mended with timbers salvaged from the wrecks.

And Caesar's problems grew worse. His cavalry had been delayed from Gaul, and when these set sail across the Channel in 18 ships they were beaten back by the same storm.

'Just as they were arriving at their destination and were in view of the camp such a bad storm suddenly arose that none of the ships could keep on course... some were carried back to the port from which they had started, others were swept down, in great peril to themselves, to the lower, that is, the more westerly part of the island. They anchored nevertheless, but when they began to fill with waves they were compelled to stand out to sea with the oncoming night, and proceeded to the Continent.'

The lack of cavalry seriously restricted Caesar's operations. Although he managed to fend off further attacks from the Britons, time was running out before the autumn equinox arrived, which the Romans believed would bring more storm weather and leave them stranded in England for the winter without provisions. 'Since the time of the equinox was near, he did not consider that, with his ships out of repair, the voyage ought to be deferred till winter.' Once the ships had been repaired, the army was evacuated and taking advantage of fair weather they set sail a little after midnight, and the whole fleet reached the mainland. And so the first Roman invasion of England ended.

Storms around the time of the autumn equinox were something that mariners had long dreaded, although it is actually a myth. The equinox itself has no impact on the weather – it is simply the time of the year when night and day are roughly equal length, and when the sun is directly overhead at noon at the equator. The legend of the equinoctial gales could have taken root because mid-September, around the time of the autumn equinox, is often the height of the Atlantic hurricane season. If a hurricane dies out in the Atlantic its warm, humid remains can sometimes become absorbed into a depression, which becomes super-charged into quite a ferocious storm that can later batter northwest Europe.

Caesar was not put off, though, and next year he returned with an even larger invasion force. But again he was thwarted by the weather and he was trapped in Gaul by unfavourable winds for four weeks. Eventually in July he set sail with some 35,000 men and 4,000 cavalry and landed on sandy beaches, possibly near Sandwich in Kent. But after landing, his fleet was again hit by a storm, and the ships ran aground or crashed into each other. Around 40 vessels were completely wrecked and the rest needed repairing. Despite this setback, though, Caesar spent two months marching through southeast England attacking the local tribes, but returned to his fleet in Kent to sail back to Gaul. So many ships had been lost that he needed to make two journeys to return his forces to Gaul. The first leg of the journey went well, but when the ships tried to return to Kent they were beaten back by more storms. Eventually Caesar feared he would be stranded in England, and packed all his remaining troops into the few remaining seaworthy boats and made his way back heavily overcrowded. This time the Channel was so calm that the crossing had to be made by rowing back to Gaul.

It was another nine years before the Romans successfully invaded and conquered England.

Battle of Hastings

1066

The year 1066 was a time of great political intrigue. Edward the Confessor died at the start of the year but left no heir, and so his right-hand man Harold, Earl of Wessex, took over the throne. However, he faced two competing claims to the throne, from Duke William of Normandy and King Harald Hardrada of Norway.

That summer, King Harold feared the Norman invasion most and stationed large numbers of Anglo-Saxon forces along the south coast waiting for the Norman fleet, which had assembled in early August on the northern coast of France. But for six weeks the Norman ships were pinned down to their anchorage by a persistent wind from the north and could not set sail.

But those same northerly winds allowed King Harald to sail from Norway and in September he invaded the north of England, quickly crushed the defending Anglo-Saxon forces and took the city of York. Learning of the Norse conquest, Harold rushed north and assembled an army on the way in a remarkable four days. The Norse invaders were not expecting another attack so soon after their victory, and on a warm day in September they were caught completely offguard as they relaxed without their armour. The Saxons routed them at Stamford Bridge, outside York, and Harald was killed.

But Harold's celebrations were short-lived. Just days later, the wind changed to southerly and now the Normans set sail. They landed easily on the undefended coast at Pevensey in Sussex on 28 September and set up camp.

Perhaps if the northerly wind had lasted longer Harold might have had enough time to reach the south of England and repel the Normans before they could establish a beach-head. But in the event, Harold now raced back from the north to tackle this new outbreak of trouble.

The two forces met at Hastings, on 14 October 1066, and taking the high ground on a slope, Harold took up defensive positions and repulsed waves of Norman attacks. But in a long and bloody battle Harold was eventually defeated. The Battle of Hastings marked the end of the Anglo-Saxon age and the last time that England was successfully conquered by a foreign power. It resulted in a new language, a feudal state, changes in law and in government, but also set in place centuries of war with France.

Winchelsea

The Atlantis of Sussex

The old town of Winchelsea near the Sussex coast is a quaint medieval place full of bric-a-brac and cream tea shops, but strangely it was a new town planned and built on a rectangular grid pattern with streets numbered rather than named. In fact, it is not the original town at all.

Old Winchelsea lay a few miles away on a large shingle bank, a major fishing and trading port famous for imports of French wines, a key gateway to Normandy, a safe anchorage for passing shipping, with a thriving shipbuilding and ship repair industry and a lucrative sideline in piracy.

But from 1233, a series of storms shattered the old town. One of the worst storms struck on 1 October 1250, described by the chronicler Raphael Holinshed: 'The sea appeared in the dark of the night to burn as if it had been on fire, and the waves to strive and fight so that the mariners could not devise how to save their ships.' Three large ships and many smaller ones sank, their wreckage strewn for miles, whilst bridges, windmills and 300 houses were swamped.

Just two years later, the writer Matthew Paris told of another storm, on 14 January 1252, that swept England.

'A terrible wind prevailed, – drove back the sea from the shore, tore off the roofs of houses, or threw them down, uprooted completely the largest trees, stripped churches of their lead, and did other great damage by land, and still greater by sea, and especially at the port of Winchelsea... the waves of the sea returned and came into the shore, and overflowed the mills and houses, and carried away a number of drowned men.'

Edward I was so concerned with Winchelsea's pummelling by storms that soon after his coronation he visited the town and realised it had to be rebuilt on another site. Plans were drawn up for a new town on a nearby hill, but before the work could be completed one last, devastating storm struck in 1287. This was so violent that the coastline was reshaped, rivers took new courses, and Old Winchelsea was wiped off the map, although its ruins appeared at low tide for years afterwards. Today the site of the old town is roughly where Pontins holiday camp lies at Camber Sands.

Duke of Buckingham's Water

1483

It came as quite a shock in the wet of summer of 2007 to discover how the Rivers Wye and Severn can devastate so much of middle England in floods. But these rivers have a long history of disastrous flooding, and in one case helped defeat a rebellion that threatened the monarchy.

In 1483, Edward IV died and left the succession to the throne with his two young sons. But Edward's brother, Richard, staged a coup d'etat, imprisoning the two boys in the Tower of London and crowning himself Richard III. In this astonishing act of treachery he was aided by Henry Stafford, Duke of Buckingham, who was richly rewarded by the new king. But no sooner was the coronation over than Buckingham planned his own breathtaking treachery and plotted a rebellion to unseat Richard.

Buckingham planned on raising forces in and around Wales, before crossing the River Severn at Gloucester and joining up with other rebel forces in the West of England. Buckingham declared his rebellion at Brecon, South Wales on 18 October, but his advance was soon stopped dead in its tracks by a fearful storm. 'In the evening there was the greatest wind ever heard of, which caused a wonderful great flood in most part of the land from Bristol to "Mount" and many other places, drowning the Counties roundabout,' recorded one chronicle of the time.

The rains were horrendous and both the River Wye in Herefordshire and the Severn in Worcestershire rose rapidly.

'There was so great an inundation of water that men were drowned in their beds, houses were overturned, children were carried about the fields swimming in cradles, beasts were drowned on the hills.' This extraordinary flood was reckoned to have drowned over 200 people and afterwards became known as 'The Duke of Buckingham's Water'.

Crossing of the Severn was impossible for ten days and left Buckingham hopelessly stranded. He had insufficient supplies, his Welsh followers deserted and the rebellion soon collapsed in disarray. Buckingham himself fled to the house of an ally near Shrewsbury, hoping to escape abroad later, but in another ironic twist he was betrayed, turned over to the king's forces and executed on 2 November.

The same storm that brought to an end Buckingham's rebellion also thwarted the return of Henry Tudor from exile in Brittany, who had his own pretensions to the throne. He was forced to return to Brittany, soon to be joined by other rebels fleeing Richard III who swept through the West Country afterwards.

But the rebellion had revealed the deep unpopularity of Richard and seriously undermined his power. Richard III's reign lasted only two years before Henry Tudor successfully returned to England and defeated and killed him at the Battle of Bosworth Field, Leicestershire. The demise of Richard brought an end to the War of the Roses, and Henry was crowned Henry VII, the first monarch of the Tudor dynasty.

Great Ball of Fire

Widecombe

The village of Widecombe nestles amongst the rolling hills of Dartmoor, with a towering church known locally as the Cathedral of the Moor. On 21 October 1638, a congregation of some 300 souls were packed inside the church for a Sunday service led by Rev George Lyde, when it became intensely dark until the worshippers could no longer read their prayer books. 'A mighty thundering was heard, the rattling whereof did answer much like unto the sound and report of many great Cannons; and terrible strange lightning therewith, greatly amazing those that heard and saw it,' wrote one chronicle. 'Extraordinary lightning came into the Church so flaming, that the whole Church was presently filled with fire and smoke, the smell whereof was very loathsome, much like unto the scent of brimstone.'

'Many others in the Church did see presently after the darkness, as it were a great ball of fire come through a window and fly around the church, apparently blasting open the roof, tearing through the wall of the church tower, rebounding "like a cannon ball" until it shot through the aisle.'

A pinnacle on the church tower collapsed and crashed through the roof, sending a large beam crashing down and ripping stones out. Many of the congregation were thrown to the ground, some dashed their heads against pillars and died, others were horribly burnt with their clothes on fire. Several people were killed and some 60 injured.

Exactly what caused so much pandemonium is intriguing. The ball of fire had all the hallmarks of a violent episode of ball lightning, a rare and mysterious form of lightning usually created in exceptionally severe thunderstorms. But it is unlikely to have created such large-scale damage on its own. Much of the structural damage to the church may have been caused by a ferocious wind. One clue came from the landlady of a nearby inn who described how the Devil himself had passed through Widecombe that day, and ordered ale that sizzled and steamed as he drank it. It is interesting that the devil often referred to tornadoes in those days, and it is possible that an exceptionally violet thunderstorm spawned both a tornado and ball lightning, although cases of this are extremely rare.

There have been other cases of violent balls of lightning, what might be called fireballs. These have been seen plunging out of thunderstorms and exploding into flames on the first thing they land on, and one terrifying incident happened on 20 July 1992, in Dormansland, Sussex. A family was sat in their conservatory watching a thunderstorm when they heard a huge explosion, the lights went out and they felt their hair stand up on end. Soon after they saw a small ball of red light shoot across the darkened living room. Meanwhile there was a commotion from outside as their neighbours had seen a 6ft (1.8m) hole blasted through their roof of the house before it was set ablaze. Eyewitnesses further away had seen a huge red ball of fiery light plunge out of the storm clouds and burst into flames on the roof. This was clearly an electrical phenomenon because all the electrical wiring in the house was burnt out, including television, video and any other equipment that had been plugged in to sockets. But beyond that it is difficult to explain this incident.

Great Fire of London

1666

England in 1665 was in crisis. The Plague reached epidemic proportions and people fled cities and towns in fear of contamination. Added to the misery, an intense drought began during the hot summer that year. As Samuel Pepys wrote on 7 June, 'it being the hottest day that ever I felt in my life, and it is confessed so by all other people the hottest they ever knew in England in the beginning of June.'

The lack of rain continued through the winter into the following year, by the dry spring in 1666 Pepys noted on 18 March, 'So walked to Westminster, very fine fair dry weather, but all cry out for lack of rain'.

The Plague also continued, and even though Charles II returned to court at Whitehall in February 1666, London remained unsafe, with death carts still commonplace – 68,000 people in London died of the disease in two years. A big fear of the inhabitants of the city was the persistent, strong east wind, along with the dry, dusty air, which they thought carried the plague. In fact, the dry winds were stoking up an entirely different disaster.

The summer of 1666 turned extremely hot and on 5 July 1666, Pepys noted: 'extremely hot ... oranges ripening in the open at Hackney.' The drought grew so bad that rivers ran almost dry, even though several thunderstorms seemed to give some relief. 'It proved the hottest night that ever I was in my life, and thundered and lightened all night long and rained hard,' Pepys wrote on 7 July.

Even Scotland was baked dry that summer, whilst in Oxford the rivers ran almost dry 'to the great impoverishment of boatmen.' John Evelyn recorded in his diary how the drought helped lead to the final calamity: 'This season, after so long and extraordinarie a drowth in August and September, as if preparatory for the dreadfull fire.'

Every month from November 1665 to September 1666 was parched, and London's ramshackle wooden buildings had turned into a tinderbox, with river warehouses full of oil, tar, pitch, hemp, flax and gunpowder, amongst other highly combustible materials. The Thames was running low and much of London's underground springs dwindled.

At 2am on Sunday, 2 September, smoke began pouring out of the house and shop of Thomas Farynor, the king's baker in Pudding Lane, near London Bridge. With only narrow streets dividing the city's wooden buildings, the fire took hold rapidly, and within an hour the Mayor, Sir Thomas Bloodworth, had been woken with the news. He was unimpressed, declaring that 'A woman might piss it out'. Yet by dawn London Bridge was burning.

Samuel Pepys lived nearby and next morning walked to the Tower of London where he saw the fire heading west, fanned by the wind, and described 'pigeons... hovering about the windows and balconies till they burned their wings and fell down'. The King ordered Bloodworth to blow up as many houses as necessary to make a firebreak to contain the fire, but these early efforts were overcome by the fierce wind, which sent flames leaping over houses. By the end of Sunday the inferno had turned into a firestorm with its own winds, as air was sucked into the flames, sending sparks and pieces of burning material flying off in different directions.

By 3 September the fire was raging north and west, and panic reigned. The Duke of York took control of efforts to stop the fire, with militias summoned from neighbouring counties to help the fight, and also to stop looting. But the flames continued relentlessly, devouring Gracechurch Street, Lombard Street, the Royal Exchange, and heading towards the wealthy area of Cheapside. By mid-afternoon the smoke could be seen from Oxford, and bits of burnt paper were carried by the wind as far as Windsor. Londoners fled to the open spaces of Moorfields and Finsbury Hill and set up refugee camps there.

On 4 September the fire was still raging, as Evelyn recorded, 'The eastern wind still more impetuously driving the flames forward.'

Eventually, on 5 September the east wind dropped by the afternoon, and the fires were burning themselves out. The Great Fire of London had raged for four days before the winds dropped and the fire lost intensity and broke up, burning down more than 13,000 houses, 87 churches and St Paul's Cathedral, some four-fifths of the city and leaving up to 200,000 people destitute and an unknown number of deaths.

The city was rebuilt in brick and stone, built on the existing streetplans to avoid disputes over land ownership, but the streets were made wider with clear access to the river, and sewers were added. After the fire the Plague ended, after some 80,000 deaths in London, possibly by burning down insanitary buildings and their rats. Given that the Plague did not reappear in London, the fire may have been an effective way of getting rid of the epidemic.

The Protestant Wind

Throughout history, England has been saved from invasion several times by the timely intervention of the weather. But on one occasion an amazing sequence of weather events actually eased the way for an invasion.

In 1685 James II succeeded to the throne, but as a Catholic he was at odds with his predominantly Protestant subjects. Things came to a head when James's wife gave birth to a son, signalling the start of a new Catholic succession. England was thrown into political crisis, and powerful establishment figures sought help from the Dutch Protestant William of Orange to take the British throne, based on the right of succession of his wife, Mary. It was an arrangement that suited William perfectly.

In 1688, William assembled a colossal invasion force larger than the Spanish Armada, some 500 ships carrying an army of 20,000 soldiers, thousands more sailors and 7,000 horses. But he needed a favourable wind and for three weeks the weather blew from the southwest, pinning down the Dutch fleet to their naval base. Time was of the essence as it was getting late in the year to launch an invasion, and by late October the Dutch were getting anxious that winter was fast approaching. As Bishop Burnet, one of William's confidantes, recalled: 'The wind had stood so long in the west, there was reason to hope that it would turn to the east; but when that should happen no time was to be lost; for it would blow so fresh in a few days as to freeze up the river, so that it would not be possible to sail all winter long.' Eventually, on 1 November the weather changed and an easterly wind allowed the Dutch to set sail into the

North Sea, whilst the English fleet was hemmed in by the wind at Harwich on the Essex coast. Then the wind changed and carried William swiftly south, past the English navy and into the Straits of Dover, where crowds stood on the coast and watched in awe as the magnificent Dutch fleet sailed past on 3 November. By the time the English navy finally set sail, William was far in the lead.

As the Dutch ships sailed past the coast of Devon, they had another stroke of luck as Bishop Burnet described: 'The wind turned into the south, and a soft and happy gale of wind carried the whole fleet in four hours time into Tor Bay.' William landed safely in the shelter of Torbay, near Brixham in Devon, and by another extraordinary stroke of luck the wind changed again with a strong westerly gale that blew James's fleet back down the Channel. Finally, conditions calmed down and allowed William's troops to land ashore safely.

All told, the wind had changed five times in five days, each time in William's favour. 'This strange ordering of the winds and seasons, just to change as our affairs required it, could not but make a deep impression on me, as well as on all that had observed it,' remarked Bishop Burnet in amazement. Small wonder this was called the Protestant Wind.

As William's invasion force slowly made its way through the muddy roads of the West Country, support for James II collapsed. The king retreated to London and when the Dutch arrived he was captured, taken out of the capital and allowed to flee to France. William and his wife Mary were crowned joint monarchs in what became known as the Glorious Revolution, an invasion that the British acquiesced to, but which had far-reaching consequences. The Protestant succession to the throne was ensured, but the Monarch never again held absolute power. Dutch influence flowed into England, its effects seen in paintings, buildings, gardens and even the Dutch national drink of gin. And Dutch banking transformed London as a commercial centre and England boomed, growing into a rich and powerful nation after 1688.

The Great Storm

1703

British storms, by and large, hardly compare to the power of tropical cyclones, typhoons and hurricanes. But on 26 November 1703, the worst storm recorded in English history was so ferocious it would qualify as a Category 3 hurricane. Nothing in recent times has come close to the ferocity, death and destruction of such a tempest. Winds raged at up to 120mph (190km/h) through the night, driving up waves 60ft (18m) high that pulverised the coastline. The storm carved a swath 100 miles (160km) wide through southern England, devastating towns and cities, and left some 8,000 people dead. Around a tenth of the Royal Navy's manpower was lost, along with some of its finest warships. Millions of trees were torn up, barns swept away and windmills burnt down. 'Never was such a storm of wind, such a hurricane and tempest known in the memory of man, nor the like to be found in the histories of England,' wrote the *Observator* newspaper.

That November was incredibly stormy as a series of Atlantic gales swept the country. But on 25 November a lull in the winds allowed ships to make for safe anchorages as ports around the coast frantically unloaded cargoes and berthed vessels, with every harbour and estuary full of merchant vessels. The storms could not have come at a worse time for the Royal Navy, either. The year before, England and its Dutch and Austrian allies had gone to war with France over the successor to the Spanish throne – the War of Spanish Succession – and after launching raids in the Caribbean, Atlantic and Mediterranean, the fleets were returning to their winter quarters at home.

This was the calm before the storm, though, and the following morning the winds picked up again. By nightfall the sky was thick with clouds and the wind was howling. Ships' barometers recorded alarming falls as a depression from the Atlantic exploded in what meteorologists today call a 'bomb' for the way its pressure suddenly plunged to an intense low. 'Between the gusts it sounded like thunder in the distance,' described one account, 'of real thunder and lightning there was none, but in some places the air was full of meteoric flashes.'

Ships at sea and at anchor rapidly took down sails ready for the worst, and as conditions rapidly deteriorated some ships even cut down their masts to prevent them being dragged by the wind. Hundreds of ships left out in the English Channel faced a horrific ordeal and many were sunk by monstrous waves, crashed into other ships barely visible in the night, or were wrecked on reefs and rocky coastlines.

Not even the best defended harbours were safe. Ships were ripped off their anchors in ports and left battered on stone walls, smashed into each other, or tossed out to sea. One small ship anchored in the River Helford in Cornwall was torn off its moorings, hurled out to sea and eight hours later ended up aground on the Isle of Wight, 275 miles (440km) away.

Buildings on the coast also bore the brunt of the storm. The Eddystone lighthouse was the world's first lighthouse built entirely offshore, on a craggy rock 14 miles (23km) off the coast from Plymouth. It had successfully weathered five years without mishap, but its architect, Henry Winstanley, boasted that he wanted to be in the lighthouse 'in the greatest storm that ever blew under the face of the heavens'. He managed to land on the lighthouse just before the storm broke and witness the storm first hand. It was not a happy experience. Towering waves crashed onto the lighthouse, shattering the timbers at its top, ripping beams apart and eventually tearing the stone base wide open. By the following day all that remained were a few bent pieces of iron sticking up from the rocks, without trace of Winstanley or the lighthouse crew.

The greatest shipping losses struck the northeast coast of Kent, where some 160 merchant ships and several men-of-war came to grief, many driven on to the Goodwin Sands, a notorious stretch of shoals and sandbars. However, in one dramatic escape *The Association*, a giant 96-gun man-of-war of the Royal Navy, was ripped off its anchorage on the Essex coast near Harwich and hurled across the North Sea all the way to Gothenburg, Sweden.

The Severn Estuary caught the full fury of the storm as a storm surge piled directly into its funnel-shaped mouth. As the mass of water rammed into the neck of the estuary it got squeezed, the sea level rose rapidly and burst through the earth bank sea defences and inundated the flatlands of Monmouth and Somerset. Thousands of sheep and cattle were washed away like flotsam, villages and farms disappeared, and even ships were swept inland on the surging waters – one vessel was dumped 15 miles (24km) inshore. The bodies of hundreds of sailors from shipwrecks were found in fields for weeks afterwards.

Further inland, the country rapidly took on the look of a battlefield. Streets were left scattered with the debris from hundreds of slates torn off roofs, crashed chimneys and fallen trees. The Bishop of Bath and Wells and his wife were both crushed to death by a falling chimney that smashed through their roof, bedroom and down onto the ground floor below. There were widespread accounts of an earthquake at the height of the storm, but this was probably caused by the fearsome winds shaking buildings so violently. Many people said that there was fearful thunder and lightning, some of it quite strange and inexplicable. Joseph Clench, an apothecary from Jermyn Street, London, reported lightning flashes 'seemed rather to skim along the surface of the ground; nor did they appear to be of the same kind with the common lightning flashes.'

London received the full onslaught of the storm at around 3am. Entire houses collapsed, others lost roofs, walls and windows. Falling chimneys made lethal bombs that crashed through densely packed buildings. Queen Anne had a close escape at St James's Palace as she watched the storm from a window when part of the roof gave way and a stack of chimneys plunged down, killing a servant. She quickly rushed to shelter in a cellar under the palace, as she later declared: 'Calamity so Dreadful and Astonishing, that the like hath not been Seen or Felt, in the Memory of any Person Living in this Our Kingdom'.

We know much about the storm thanks to Daniel Defoe, author of *Robinson Crusoe*. On the night of the storm, Defoe was staying close to London, possibly Newington Green, and almost lost his life when a chimney smashed to the ground in the street close to him. Next morning he could not believe his eyes when he ventured outdoors after the storm: 'Streets covered with Tyle-sherds, and Heaps of Rubbish, from the Tops of the Houses, lying almost at every Door... the Distraction and Fury of the Night was visible in the Faces of the People'.

He described extraordinary scenes of destruction on the river: 'some vessels lay heeling off with the bow of another ship over her waste, and the stern of another upon her forecastle. ... some lay so leaning upon others, that the undermost vessels would sink before the other could float.' Virtually all 700 ships moored in the lower reaches of the River Thames were blown into a colossal heap of wreckage around Limehouse.

Defoe wanted a sense of the national scale of the disaster, something never achieved before in weather history. He advertised in the *London Gazette* for eyewitness reports from all over the country, publishing them in a volume entitled *The Storm* in 1704. In this he noted how church spires were toppled; in one case the Rev Samuel Farr, in Stowmarket, Suffolk, described how the spire of his church was blown 'clean off', thrown 28ft (8.5m) down the length of the church before smashing through the roof. The wind was so strong it stripped church roofs of their lead and Defoe described how the lead roofing on Westminster Abbey 'rolled up like parchment and was blown clear of the building'.

He estimated that 400 windmills were destroyed as their wooden brakes broke and their sails spun out of control so furiously that the intense friction set the mills ablaze.

At least 123 people were killed on land apart from thousands lost at sea, but these are only rough estimates and the true figures were possibly much higher. Considering that the population of England was only about 5 million, this was a huge loss, and by far the deadliest storm in recorded English history.

The economic consequences of the storm were soon felt. The cost of roofing materials soared, the price of tiles rose from 21 shillings to 6 pounds per thousand. Ironically, Defoe had lost his own tile business the year before. Labourers profited from the disaster by inflating their costs, bricklayers doubling their fee to 5 shillings a day. In fact, costs became so exorbitant that many householders were forced to repair their roofs with wooden boards, thatch or twigs.

The price of food also rocketed because so many farm barns full of grain were destroyed, so the price of wheat, hay and straw almost doubled.

The Royal Navy lost around 3,000 men and there were fears that the French would take advantage of the weakened English and launch an invasion. What no one realised was that the French had suffered their own shipping losses along the Channel, and posed no threat. But the Navy, which was always short of men, faced an enormous recruitment problem to replace its manpower losses, particularly as the merchant fleet had suffered so badly as well. As a result, the House of Commons agreed that foreign nationals could be allowed to serve in the Navy, and even prisoners of war.

Another problem was the huge demand for skilled workers to repair the damaged or build new ships. One consequence was to make ships far less ornate with far fewer wooden carvings to speed up their construction. However, the dockyards responded with a phenomenal rate of work and by the following year the Navy was back up to strength. In fact, Sir George Rooke scored a stunning victory by capturing Gibraltar, one of the most spectacular conquests of the entire war with France.

The landscape also changed. The diarist John Evelyn was deeply upset at the sight of thousands of fallen trees. 'So many thousands of goodly Oaks subverted by that dreadful Hurricane. ... lying in ghastly Postures, like whole Regiments fallen in Battle.' Millions of trees were felled by the storm, but it is unlikely that the fallen timber went to waste with so much reconstruction to be done.

What Caused The Great Storm?

Defoe himself made some very astute scientific observations, suggesting that the storm may have begun as a hurricane that he thought had struck Florida a few days beforehand and its remnants then crossed the Atlantic.

But this was the wrong time of year for hurricanes and there were no records of tropical storms even weeks beforehand. Weather historian Dennis Wheeler at Sunderland University searched dozens of ships' logs from the Caribbean and Florida at the time and found no mention of a hurricane around that time, although they do mention warm conditions. However, further north he discovered that HMS *Centurion* sailing off the coast of Boston recorded atrocious snow, ice and freezing gales. So it seems possible that bitterly cold Arctic air had erupted into the Atlantic and collided with warm, tropical air from the south, and this could have set off the Great Storm.

Robert Muir-Wood at Risk Management Solutions (RMS), a company that calculates weather risks for the insurance industry, holds up the storm of 1703 storm as an example of how badly prepared Britain is for such extreme events. He has studied the pattern of damage from that storm and calculates that it packed sustained wind speeds well over 100mph (160km/h), far more powerful than the October 1987 storm. If such a storm struck today he estimates that damage could be up to £15 billion. Millions of people would be left without power, transport would be in chaos, and if it struck during daytime in the week it could cause hundreds or maybe thousands of deaths. The worst hit places may be school buildings, many of which are poorly built, as well as caravan parks and vehicles on the road. 'It would be the biggest natural disaster that Britain has ever had,' Muir-Wood warns. Perhaps the lessons of 1703 have not been fully appreciated yet.

The Margate Desert

1921

In 1921 England was gripped by an extraordinarily prolonged drought. High pressure from the Azores remained stuck for almost the entire year, in what was the most anti-cyclonic year since records of pressure began in 1861.

Kent was hit especially hard, and the countryside turned into a parched yellow landscape shimmering in heat hazes, more like the Mediterranean in the height of summer. Margate was hit worst of all, with a mere 9.29in (236mm) rainfall over the entire year, a record for the lowest ever yearly rainfall anywhere in the UK. In fact, it technically qualified as a desert, defined as a year's average rainfall less than 10in (255mm) precipitation.

That heat and drought may have inspired T.S. Elliott to write the poem 'The Wasteland', which has been widely interpreted as reflecting the First World War. But decades later the original manuscript turned up in a pile of papers, amongst which were hotel receipts from Margate dated 1921, showing that Elliott had stayed there whilst writing the poem, and maybe was inspired by the desert-like conditions of Margate.

In one audacious experiment, an attempt was made to break the drought using fireworks on Hampstead Heath, London. The firework company C.T. Brock claimed explosions from a huge fireworks display could somehow disturb the atmosphere and trigger rain. London County Council granted permission to make a trial run, and one fine warm evening in July, thousands of people, including a man from the Meteorological Office, gathered to watch the spectacle.

'Hundreds of rockets and aerial bombs were fired in rapid succession, and the clear blue sky was immediately speckled with tiny puffs of smoke,' reported the *Hampstead and Highgate Express*. 'But of rain there was not a drop. There was not an umbrella among the crowd and even the organisers of the experiments had forgotten to bring their trench coats.'

The skies over London remained stubbornly clear and dry for weeks afterwards. In fact, hardly a drop of rain fell until mid-September. It was the best season for Britain's holiday resorts and August Bank Holiday was reported the most prosperous. 'All the bathing stations were crowded and the seashore presented scenes of great animation,' gasped the *Margate Gazette*.

Even October carried on as if summer had never ended, thanks to persistent anti-cyclones. It was the sunniest October on record and crowds flocked to the seaside in temperatures hitting the high 20s Celsius (80s F), with the *Daily Telegraph* on 21 October triumphantly declaring, 'No hotel rooms were available within a 50 mile radius of Blackpool and holidaymakers were forced to camp in tram shelters on the promenade.' Temperatures peaked at 29°C (84°F) on 5 and 6 October in London where *The Times* noted incredulously, 'Two or three men, it's true, were detected without their waistcoats.'

By then the country was in chronic drought. The annual ploughing match at Margate had to be postponed because the ground was so hard, and as the *Isle of Thanet Gazette* remarked, 'Never before has Margate enjoyed such a summer of sunshine... It will go down in history as the wonder-year of weather.' Over the whole of England and Wales, it was the second driest year on record.

However, the north of Scotland did not share in the heat, though, and in early October it broke another record, when a coastguard officer reported two large icebergs about 30 miles (48km) north of the Butt of Lewis. 'Not within living memory have icebergs been seen in these waters, and their proximity goes far to explain the low temperatures previously lately in the North of Scotland,' observed the *Daily Telegraph*.

The Cold and Frosty Climate of Rickmansworth

Rickmansworth in Hertfordshire is a genteel commuter town at the end of the Metropolitan Underground line, 17 miles (27km) outside central London, and yet it has a climate of extreme cold remarkably similar to Braemar in the Scottish Highlands, one of the coldest inhabited regions in the British Isles. The average minimum for Rickmansworth from 1906–35 was only 0.5°C warmer than Braemar. As a result, gardeners in Rickmansworth can expect dahlias to be lost to frost in mid-August, while in neighbouring areas they can survive until November. And yet on 29 August 1936, Rickmansworth also recorded the biggest swing in a day's temperature in the UK, when the temperature swung by 28.3°C (50.9°F) in just nine hours. The day began cold and foggy, but once the fog cleared the temperature soared from 1.1°C (34.0°F) at dawn to 29.4°C (84.9°F) by mid-afternoon. These extremes probably make Rickmansworth one of the most continental-type climates of Britain.

What makes Rickmansworth's climate so bizarre? The weather station there was sited at the bottom of a deep sheltered valley at the foot of the Chiltern Hills. On a calm, clear night, cold air can spill down the steep valley slopes like running water, and sometimes can actually be seen. 'On quiet, clear autumn evenings when garden bonfires of damp leaves are burning, it is common to see during the hour after sunset, rivers of white smoke slowly winding their way down into the northern strip of the valley,' wrote E.L. Hawke, resident of

Rickmansworth and Secretary of the Royal Meteorological Society. The cold air is then trapped on the valley floor by a high railway embankment across the valley, and so makes an ideal frost hollow. But the valley slopes also make a suntrap that heat up during the day under the sun.

Other well-known frost hollows in the UK are the Welsh Marches, the glens of Scotland, the Pennine valleys, the Vale of Evesham, Shrewsbury and Redhill. Frosts are often seen here earlier in the autumn and later in the spring than on the surrounding higher land. And you can often see a frost hollow when they are filled with a shallow pool of ground fog.

But weather stations in frost hollows have been closed down because they are not thought to be representative enough of the country's climate. It was a problem that E.L. Hawke lamented in the 1930s, 'Hence opportunities for determining the "coldest places" are deplorably poor,' he wrote.

Blue Moon

On 26 September 1950 an astonishing sight brought much of Scotland gasping. The Sun turned blue.

'There it was, completely reversing the normal state of matters – a blue sun shining in a white sky,' described the *Scotsman* in Edinburgh. 'All over the city people stopped to gaze at the sapphire sphere.'

'For about an hour telephone lines to newspaper offices all over Scotland including the office of *The Scotsman*, were jammed by inquirers... there were actually some persons in Scotland who voiced the fear that the end of the world was coming.'

For good measure there was a blue moon that evening as well.

An RAF Meteor jet fighter was sent up over Scotland to investigate and reported that the Sun was a vivid blue up to 5 miles (9km) high. Above that was a smoke layer of brown haze, and at 8 miles (13km) altitude the Sun looked normal.

Meteorologists suspected that smut was involved. Smoke from a giant forest fire raging in Canada billowed up high into the atmosphere and skies across northeast America turned pink or yellow with a purple or orange Sun. That smoke was then seen drifting out across the North Atlantic and the particles of smut were just the right size to scatter all the wavelengths of sunlight apart from blue, making the Sun and Moon turn blue.

Another episode of blue suns and moons happened shortly after Krakatoa erupted, on 10 December 1883. The eruption of the volcano was the loudest explosion in history, with a shockwave that registered on barometers in London. The blast of ash was shot high into the stratosphere and swept around the globe in a veil of dust that screened out sunlight, cooled the world and sent weather patterns into chaos.

The dust also created spectacular light displays in the sky. A few days after the eruption, the Sun and Moon turned blue or green in many places. In New York the sky lit up into such a brilliant scarlet one evening that the fire service was inunundated with calls from people fearing the city was on fire. In London, meteorologist George Symons reported: 'We saw a green sun, and such a green as we have never seen, either before or since.' The artist Edvard Munch described a setting sun near Oslo soon after the eruption: 'Clouds like blood and tongues of fire hung above the blue-black fjord and the city... I felt a great unending scream piercing through nature.' That surreal experience may well have inspired his famous painting 'The Scream'.

In London, the landscape painter William Ascroft made crayon sketches in Chelsea of intense blood-red sunsets and intense twilight afterglows seen long after sunset. The following spring, the afterglows began to fade and a new phenomenon, the 'Bishop's Ring', appeared in the skies over Europe. Ascroft's work showed the Sun wrapped in multi-coloured rings of light, bluish white in the centre, shading off into a reddish brown border on the outside.

And that was not the end of the strange phenomena following the eruption of Krakatoa. In the summers of 1884 and 1885 Ascroft described 'bright lights on clouds very late at night' and his sketches show these clearly as noctilucent clouds, the first known pictures of these clouds. A number of observers incorrectly referred to them as aurora, but these eerie ripples of shimmering silvery electric-blue clouds lie in the upper atmosphere, some 50 miles (80km) high, on the edge of space, where it is a thousand times drier than the Sahara and temperatures drop to −148°C (−235°F).

The years after Krakatoa erupted were the first records of noctilucent clouds but how the eruption caused them remains uncertain. Noctilucent clouds remained rare and elusive until relatively recently, when they are growing more common and more widespread. These recent sightings are nothing to do with volcanic eruptions, but seem to be a sign of Earth's changing climate as carbon dioxide seeps higher into the upper atmosphere, where, paradoxically, temperatures are growing colder whilst the lower atmosphere is warming up. So these strange clouds may actually be a very visible sign of climate change.

The Collapse of Ferrybridge
1965

The colossal cooling towers at power stations look like mountains that could never been shifted. But engineers were appalled when Ferrybridge Power Station, near Pontefract in Yorkshire suffered the unthinkable on 1 November 1965.

Cooling towers are empty tubes in which hot water from the power station falls from the top onto slats below to cool it enough to reuse. The rest goes out the top of the tower as steam.

Ferrybridge had eight cooling towers arranged in two rows, and were truly massive, wider than any previous towers. On 1 November 1965 a gale blew across the line of towers with strong winds, although not particularly excessive, at around 45mph (70km/h), with gusts thought to be around 85mph (135km/h).

The superintendent of the power station was working in his office when he had a phonecall to say that one of the cooling towers had fallen down. His immediate reaction was 'you're joking', but very soon afterwards there was a warning that another tower was about to collapse, and ten minutes later it did.

And 40 minutes after that, a third cooling tower began to wobble like jelly. A hole appeared in the side, the top slumped forward and the sides caved in until the entire structure collapsed. The other five towers remained standing but were all severely cracked.

When Ferrybridge was designed, a model of the cooling tower was tested in a wind tunnel, but not all eight towers together. The problem was that the towers were packed so tight together that the wind was squeezed past them like water racing through narrow rapids. This hugely accelerated the wind speed and massive turbulence created around the towers simply had not been factored into the power station design. The message was clear, that wind tunnel tests on new structures needed to simulate the surrounding buildings and landscape. And wind speeds needed to be calculated to take sudden gusts and higher wind speeds at the top of the towers into account. Engineers were astonished that 86 years after the Tay Bridge storm disaster, they still had not fully appreciated the power of the wind.

Motorway Madness

In 1964, police were called out to a multiple vehicle accident in fog on the M1 and reported that, 'As we approached the accident in fog with blue light going and both of us hanging out of the car windows trying to slow traffic down, cars overtook us at 60 m.p.h. on both sides. We could hear them crunching into the wreckage ahead.'

This was the phenomenon that became known as 'motorway madness'. After several high profile motorway pile-ups in fog, the Government in December 1965 introduced the 70mph speed limit on motorways and all other unrestricted roads, the first speed limit in Britain since 1935. It was supposed to be a temporary emergency measure, but at a heated press conference the Minister of Transport, Tom Fraser, admitted that it could be the beginning of a permanent overall speed limit in Britain. And so it proved to be.

The speed limit made no difference, though. On 13 September 1971, 200 vehicles crashed on the M6 at Thelwall, Cheshire. Five vehicles burst into flames, ten people were killed and 70 injured, the worst accident on British roads. As one ambulanceman said at the time: 'As soon as we managed to free one lot there were cries and shouts from other cars. It was just like a battlefield: wrecked cars and lorries everywhere.'

There are many theories for motorway madness, such as changes in a driver's perception, disorientation, loss of sense of speed and distance in fog. New solutions were desperately sought for the blight of fog on motorways, including metal barriers, overhead lighting, overhead hazard warning signs, but still the multiple crashes in fog carried on.

But one good thing is that fog is now far less common than 50 years ago, since the Clean Air Acts began in 1956 and cleaned up coal smoke pollution, and the incidence of fog has just about halved on average over half a century.

Fogs in the steam age were a huge factor in some of the most horrendous accidents on the railways. On 8 October 1952, during the morning rush hour, a Perth to Euston express train missed a caution signal in the fog near Harrow and Wealdstone station, north London, and ploughed into the back of a commuter train in the station, hurling wreckage across the tracks. Seconds later, a Euston to Manchester express train hit the wreckage strewn across the tracks. There were about 1,000 passengers on all three trains, and 112 people were killed and around 350 injured, some of them waiting on the platform as carriages and engines ploughed across the station, or walking across a footbridge over the tracks that was ripped apart by coaches hurled up to 30ft (9m) high. It was the second worst railway accident in British history.

In response, British Rail developed the Advanced Warning System (AWS) to sound a horn in the driver's cab as a yellow or red signal was approached. But AWS was not installed universally and only five years later, on 4 December 1957, a steam train bound for Ramsgate, Kent carrying several hundred commuters and Christmas shoppers missed a red signal near Lewisham, southeast London. The train ploughed into the back of a stationary electric train with 1,500 passengers on board, crushing the carriages and smashing into a column supporting an overhead rail bridge, which collapsed onto the crash scene. Some 90 people were killed and 173 injured, the third worst rail accident in Britain.

Shipping has also suffered huge casualties in fog. During the First World War, one of the worst shipping disasters in British waters happened in fog. RMS *Mendi* had sailed from Cape Town carrying 823 South African troops, due to be sent to the Western Front in France. The *Mendi* was being escorted by a destroyer through the English Channel on a dark and foggy early morning on 21 February 1917. As the ships rounded the Isle of Wight, the *Mendi* was rammed by the *Darro*, a large liner, almost three times its tonnage, which was steaming at full speed, without any warning fog signals. The *Mendi* was almost torn in half, keeled over and sank in less than 20 minutes. The escorting destroyer, HMS *Brisk*, lowered lifeboats and tried to find survivors, but the fog, cold and darkness prevented much of a rescue. Some 646 men died in the icy waters of the Channel, South Africa's worst single disaster of the First World War. An inquiry into the accident found Captain H.W. Stump of the *Darro* responsible for travelling too fast, failing to sound fog warnings, and also failing to give assistance to the survivors of the *Mendi*. Incredibly, Stump's only punishment was a year's suspension of his captain's licence.

However, fog has caused surprisingly few air accidents in the UK. In fact, fog inspired one of the most ingenious ways of trying to modify the weather during the Second World War when fogbound airfields were a huge menace for aircraft trying to land after bombing raids. A project was begun called FIDO, standing for Fog Intensive Dispersal Of, using burning drums of kerosene along the sides of runways to disperse the fog by heating and evaporation. It allowed the successful landing of more than 2,000 aircraft, but was too expensive and dangerous to use in peacetime.

The 1987 Storm

For an island that faces the brunt of Atlantic weather, big storms in the UK are not that unusual. But occasionally, just sometimes, a storm is so momentous that it is remembered just from its date alone. The October 1987 storm became a legend, the most devastating natural catastrophe of recent times in England, and a reminder of how savage our weather can be.

The statistics were awesome: winds gusted to over 100mph (160km/h) in many places, felled some 19 million trees, 90 per cent of Kent's roads were blocked with fallen trees, some three million households and businesses were left without electricity, 150,000 telephones were cut off and 19 people were killed.

The storm rapidly developed late on the evening of 15 October, moving quickly up from the Normandy coast. The English Channel was soon turned into a maelstrom, waves grew into monstrous peaks with crests licked into horizontal sheets of spray. The *St Christopher* cross-channel ferry spent several hours tossed about like a toy on the furious water, but survived. However, her sister ship, the *Hengist*, was less fortunate. She was in Folkestone harbour when the ship's moorings broke. Faced with being smashed onto the harbour walls, Captain Sid Bridgewater was forced to sail the ship out to sea. Soon after setting out of harbour, though, the ship lost electrical power, the engines cut out and the vessel drifted helplessly. 'I thought we were going to capsize,' admitted Captain Bridgewater. 'It was a hopeless situation, thrown around in the sea in the dark without engines, risking the lives of the 25 crew members.' The *Hengist* sent out Mayday calls but nothing could be done to save her. Within half an hour the ship crashed on a concrete bank on the shore, ripping open the hull with a hole as large as a double-decker bus. The vessel was only saved from sinking because she was beached at high tide, leaving her high and dry.

There were other miraculous escapes. At Merrow, near Guildford in Surrey, the Rev Harry Fordham had an astonishing ordeal when his house collapsed. 'There was a huge crash and I woke up on the ground floor instead of the bedroom upstairs, covered in rubble. Amazingly I was lying next to the 'phone, so I called for an ambulance and immediately afterwards the line went dead.' The message got through, though, and an ambulance managed to rush both Rev Fordham and his wife to hospital; shortly afterwards the roads became completely blocked with fallen trees.

On the hills of the South Downs overlooking Brighton, a pair of fine old windmills called Jack and Jill were badly battered. Windmills in high winds need to have their sails locked using large brakes to prevent them spinning to destruction. But the wind that night broke the brakes on the sails of Jill, which spun so violently that the friction set the mill ablaze. As smoke began pouring out, volunteers managed to crawl up the hill through the wind to form a human chain and pass buckets of water to put out the fire. 'The wind was so strong I couldn't stand up, and had to crawl up the hill on my hands and knees,' Simon Potts of the windmills' preservation society described. 'When I got inside the mill it was shaking backwards and forwards, like being onboard a ship. The top floor was full of smoke, it was pitch dark, the noise was indescribable, all you could see was the glow from the fire. It was complete panic.' The mill was saved from completely burning down but it took two years to restore the damage.

By 5am, the storm reached eastern England. The winds were so powerful they pushed back the oncoming tide, leaving mudflats and shorelines exposed that would normally be under deep water. At the port of Harwich, a disused ferry used by the government to house asylum seekers broke from its moorings and began to drift dangerously. Amazingly no one was injured and the vessel came to no harm, but there was a public outcry that led to the ship being withdrawn.

Perhaps the storm is best known for the enormous destruction to trees that night. All over the south of England the landscape was changed as trees fell like matchsticks. Barbara Willard in her book, *The Forest: Ashdown in East Sussex* (Sweethaws Press, Paperback, 1989), described the night in Ashdown Forest: 'Gathered birds were flung with panic, crying about the sky, lit by the perpetual livid flashing of pylons sagging and swaying under the weight of fallen trees,' she wrote. 'A burning salt-heavy atmosphere began to shrivel and destroy leaves. Light came to show the Forests's face, twisted and distorted by this mightiest of strokes.' Around a quarter of the forest's 2,000 acres (810ha) of woodlands were flattened.

Across the country, an estimated 19 million trees were felled, in patterns that betrayed the direction of the winds. The scale of destruction was made worse because the trees were still in full leaf, which acted as sails that toppled the trees over. And the ground was saturated after days of rain, which made it easier to uproot the trees.

But one lesson from the 1987 storm was that woodlands are far more resilient to storms than had been supposed.

Woodlands that were left untouched revealed astonishing powers of regeneration, as trees sprouted new growth from decapitated stumps. The following spring saw a huge bloom of woodland flowers such as bluebells and primroses, as woodland floors were opened up to daylight. Most animals had survived the storm surprisingly well and with plenty of wood-feeding insects feeding on the fallen timber, populations of birds and bats flourished, whilst larger mammals, such as deer, thrived on the shelter from the fallen logs. In stark contrast, man-made conifer plantations suffered huge damage, and highlighted the dangers of growing species out of their normal habitat in regimented patterns. Another problem was caused to all woodlands by moving heavy machinery to remove fallen logs, ripping up the topsoil so severely that many woodlands were left permanently scarred.

The cost of the storm was astronomical. Households in England filed 1.2 million insurance claims, and the total insurance bill came to over £1.5 billion, which at that time was the most expensive single natural disaster in the world. By macabre coincidence, the day before the storm Wall Street suffered a stock market crash and the following Monday the London Stock Market had its own infamous Black Monday share crash. It seemed like the end of the world had come.

The Weather

The wind speeds broke weather records in many locations – the strongest wind that night reached 116mph (186km/h) on the Sussex coast. The wind was so fierce that Croydon in Surrey, was covered in salt spray blown in from English Channel 40 miles (65km) away, encrusting south-facing windows and trees with something that looked like icing sugar.

But the big question everyone wanted to know was why hadn't the forecasters given any warning. It was all summed up by Michael Fish's spectacular gaffe on a television weather forecast earlier that day: 'Earlier on today a lady rang the BBC and said she'd heard that there was a hurricane on the way. Well don't worry if you're watching, there isn't.' A few hours later the storm struck. However, he had qualified his statement by saying it would be very windy and wet later on that night. 'I did broadcast saying 'batten down the hatches there's some really stormy weather on the way', but the media took it out of context,' Michael Fish later remarked.

The atmosphere over the Atlantic that autumn was strange. Cold air from the Arctic wastes of Greenland surged much further south than normal and hit warm air from the tropics, as well as the Atlantic being warmer than normal. The collision of those cold and hot air masses fuelled a deepening depression with strengthening winds.

When the depression arrived in the Bay of Biscay it parked for a while and could have passed off like many other typical autumnal storms. But it suddenly exploded into a 'bomb' – an explosive deepening in pressure that drove the high winds. The detonation may have been provided by the leftovers of Hurricane Floyd, which had hit the Caribbean and Florida. After the hurricane died out in the mid-Atlantic it left behind a mass of exceptionally warm and very moist air that gave a massive boost to the jet stream, a ribbon of wind that circles the world a few miles high in the atmosphere. The jet stream surged at speeds approaching 200mph (320km/h), and when this so-called 'jet streak' hit the Bay of Biscay, it dragged the storm at great speed towards England and hugely intensified its winds.

That evening the depression exploded, and by the time it reached the English coast shortly after midnight, it was the deepest low recorded in October for more than 150 years. The pressure recorded at various locations suddenly nosedived and then surged up again, and at Portland Naval Base shot up 25.4 millibars, the largest three-hour rise ever recorded in Britain.

Another bizarre twist in the 1987 storm was its amazing heat. For anyone out in the wind that night it was unforgettable, like being blown by a fan heater. At Heathrow Airport the temperature soared from 9°C at 6pm to 17°C by midnight, an extraordinary rise. But although south and east England felt the blast of hot air, the western side of the country was actually shivering in the cold Arctic air that was chasing closely behind in the depression.

The Post Mortem

The 1987 storm was such a disaster for the Met Office an enquiry was immediately launched into it afterwards. The weather forecasters were blamed for following their computer models too slavishly, which were prone to errors during rapid storm development where there is a lack of good observations. To compound their misery, the French forecasters had made a good prediction but their warnings were ignored by the British. Another problem revealed was a worrying shortage of sea weather observations. Of the eight weather ships that had originally been in the eastern Atlantic, six had been withdrawn from service, including a French weather ship to the north west of Finisterre, at exactly the location where the 1987 storm had brewed up.

Years afterwards, meteorologist re-examining satellite pictures of the storm discovered an unusual feature – a clear path carved straight through the storm clouds. This led to the discovery of a new and highly dangerous phenomenon called the sting jet. This was a huge burst of wind created by a surge of very dry, cold air some 3 miles (5km) high in the sky that suddenly plunged through the clouds. Rather like an avalanche rushing down a mountainside, the cold, dense air accelerated down until it crashed into the ground with winds reaching up to 100mph (160km/h). Sting jets have been discovered in other highly destructive north European storms and are thought to cause an estimated £600 million damage across Europe each year, on average.

At the time, it seemed that the October 1987 storm was a freak event. But on the 25 January 1990 another extraordinarily powerful storm battered Britain. Although the number of trees felled was far fewer, 47 people were killed because the storm hit during the daytime.

Two such savage storms in such a short period were taken as a warning sign of climate change. Sea temperatures were unusually warm around the UK in both the 1987 and 1990 storms, and it could be that heat helped to fuel the ferocious winds. It is thought that as global warming increases, Atlantic storms may grow more powerful.

WINTER

Lost Town of Dunwich Storm
January 1286

Dunwich is a quaint old village on the Suffolk coast, nestling amongst sanddunes and sandy cliffs against the North Sea. Yet in medieval times this was one of England's most important towns, a thriving port with an international trade stretching from Iceland to France. Edward I had warships built there and a huge fishing fleet landed their catch at the harbour. The town boasted a mint, eight churches, two hospitals, several monasteries and the maze of streets would have been filled with bartering and merchants.

The key to Dunwich's wealth was a shingle bank across the estuary that protected the harbour entrance. But it was a constant battle keeping the port open against currents and storms tearing into the soft, sandy coastline. Dunwich's slow decline was described by the early 20th-century writer J. Cuming Walters in his book 'Cities Beneath the Sea': 'the devastation had been great in the time of the Conqueror, but the sea continued its ravages until, in the time of Henry III, it was found imperative to check the inroads by fences. The king granted £200 towards the defences of the ancient town. On January 1, 1286, a violent storm overthrew several churches, swept away houses, and deluged part of the land.'

Another great storm in 1328 hurled so much shingle on-shore that the 'harbour was utterly choked up'. The nearby rivers carved out new courses to the sea and within a decade Dunwich was almost dead as a port, forcing out the merchant traders, fishermen and shipbuilders. By the 16th and 17th century, storms were ripping into the heart of the town itself and houses were regularly disappearing into the sea.

Today, the ghost of old Dunwich lies underwater up to a mile offshore. All that remains on land is a picturesque village perched on a sandy cliff, but its fate has grown more precarious. The rate of sea erosion has picked up speed in recent 1 as the shoreline retreats at an average of about 1m a year as storms, rising sea levels and the sinking land all take their toll. It is only a matter of time before its last remains disappear.

The Grote Mandrenke

1362

It was called the Grote Mandrenke, the Great Drowning, probably the greatest North Sea flood disaster in history, with more than half the population of the coastlines along the coast of Jutland and Schleswig drowned, and some parishes swallowed up in the sea and lost forever. The flood defence dykes were not repaired for half a century. Chronicles tell of 100,000 deaths, and modern estimates range down to 25,000 – but it was certainly a catastrophic flood.

The storm swept up from the southwest, tearing through Dublin so violently that the roofs of houses were blown to shreds before battering southern England and on into the Netherlands, Denmark and northern Germany. As the winds blew they piled up a huge bulge of water, a storm surge, which battered the coast of Europe.

Chronicles from the time speak of uprooting of hundreds of trees, collapsing houses, towers, monasteries, belfries, steeples, orchards and woods and men being choked by the wind. Windmills were blown down. Coming after the second outbreak of Plague, the storm was taken as a sign of the apocalypse.

A ferocious storm swept across England, as one chronicler recorded: 'A strong gale blew from the north so violently for a day and night that it flattened trees, mills, houses and a great many church towers.' Amongst the damage, Norwich Cathedral lost its wooden spire, sending it crashing through the roof of the Cathedral's east end. The Norman clerestory was totally destroyed.

Salisbury Cathedral was so badly damaged the Bishop of Salisbury's appeal to the Pope for funds to repair the damage states 'the building is riven, the bell tower mostly fallen'.

St Albans Abbey was destroyed and was rebuilt three years later.

The storm destroyed St Mary Matfelon in Stepney, a landmark built of white stone known as 'the white chapel of St Mary Matfelon', which gave its name to 'Whitechapel', the fine belfry of the Austin Friars at London,

The damage was so considerable that Edward III made a Royal Ordinance in the aftermath laying out punishments for labourers and merchants who exploited the reconstruction.

'under pretext of the tempest (fn. 2) of wind which has of

late unhappily occurred in divers parts of our realm, by reason whereof many buildings have been levelled with the ground, and many dilapidated, broken, and damaged, and great multitudes of tiles and other coverings have been wholly or for the greater part torn from the roofs thereof; those who have tiles to sell, and other things suitable for roofing such houses, do sell the same, entirely at their own pleasure, at a much higher price than heretofore they were wont to do; and that the tilers and other roofers of buildings, seeing so great an urgency for persons of their calling, hesitate to follow their trade, or to do any work, unless they receive excessive wages for the same, and in like manner refuse, to the no small loss and grievance of our commonwealth;—by reason thereof, by advice of our Council, we have ordered that tiles and other things requisite for the roofing of buildings shall be sold at the same price at which they used to be sold before the Feast of our Lord's Nativity last past, and at no higher rate.'

'to the contrary thereof, will take and imprison without delay, and their goods and chattels, as being forfeited unto us, will arrest, and under arrest detain, until we shall think proper to give other orders as to the same; you certifying us from time to time in our Chancery as to all that you shall do herein. Witness myself, at Westminster, the 28th day of March, in the 36th year of our reign.'

An immense storm surge in the North Sea wiped out the port of Ravenser Odd, on what is now Spurn Head, Yorkshire, at the mouth of the Humber Estuary.

'It began about evensong time in the south, and that with such force that it overthrew and brew down strong and mighty buildings as towers, steeples, houses, and chimneys. This outrageous wind continued thus for the space of six or seven days, whereby even those buildings that were not overthrown and broken down, were yet so shaken that they without repairing were not able long to stand. After this followed a very wet season, namely in the summer time and harvest, so that much corn and hay were lost and spoiled, for want of seasonable weather to gather in the same,'

recorded Holinshed in the *Chronicles of England, Scotland and Ireland*. Its fate was eulogised in the *Chronicle of Meaux*: 'That town of Ravenser Odd was an exceedingly famous borough, devoted to merchandise, as well as having many fisheries, most abundantly furnished with ships and burgesses. But ... by wrong doing on the sea, by its wicked works and piracies, it provoked the wrath of God against its self beyond measure.' Thus ended the days of Odd, in ignominy and shame, to be almost forgotten by time, history and the memories of men.

In Europe, the flood swamped Danish, German and Dutch coastlines and changed the shape of the land. It smashed the German coast into islands. In the Netherlands, the storm surge carved out a huge inland sea, what is now the Zuider Zee, connected to the North Sea and around which many fishing villages have developed.

The disaster became known as the Great Mandrenke, Great Man Drowning, and claimed an estimated 25,000 to 100,000 lives. The north German island of Strand and major port of

Rungholt in Northern Friesland, Germany, was sank almost without trace. But in the 1920 and 1930s remains of the city were exposed and relics were found until the late 20th century. Legend has it that one can still hear the church bells of Rungholt ringing when sailing through the area on a stormy night.

Killer Wave

1607

On the morning of 20 January 1607, a great wall of water surged up the Bristol Channel into the Severn Estuary. 'Huge and mighty hills of water were seen tumbling over one another to the astonishment and horror of those who saw the spectacle. Many, at first mistaking it for a great mist or fog, did not seek to escape, but on its nearer approach, which seemed faster than the birds could fly, they saw that it was the violence of the waters which had broken bounds and were pouring in to deluge the whole land.'

An area of 200sq miles (518sq km) across South Wales and Somerset were flooded, some 2,000 people drowned, and at least 30 villages were wiped out. 'So swift were these waters that in less than five hours most parts of the counties on the Severn's banks were under water, and many persons drowned. From the hills could be seen herds and flocks, all swept away together with husbandmen from the fields in one dreadful inundation; the waters were huge like to the main ocean, with the tops of churches and steeples like the tops of rocks buried in the sea.'

There is speculation that the flood was a tsunami – a giant wave born out of earthquakes or volcanoes. According to a BBC2 'Timewatch' programme shown in 2005, a tsunami up to 32ft (10m) high swamped the low-lying land around the Severn in South Wales and Somerset. The waters were said to have rushed inland, catching people unawares, advancing at a speed 'faster than a greyhound can run' and reached 25ft (7.5m) high. In the low-lying Somerset levels, it reached 14 miles (22km) inland. Experts believe a repeat flood today would cost £13bn.

The television programme suggested that the tsunami was triggered by an earthquake off southwest Ireland. But oceanographers at the Proudman Oceanographic Laboratory, Liverpool say there is no record of a tsunami anywhere else in striking distance. And parts of eastern England were also flooded around the same time, even though a tsunami from Ireland could not have reached there.

More likely, the disaster was caused by a storm surge riding on an exceptional high tide. A storm's strong winds and low atmospheric pressure can drive a huge mass of water into shallow coastlines. The mean spring tidal range of the Bristol Channel is 40ft (12.2m), the second highest in the world, making the Bristol Channel and its surrounding regions highly susceptible to flooding. Large parts of the Somerset levels are at or below sea level and these traditional flood plains routinely suffer inundation. The flooding of Somerset and Monmouthshire in January 1607 is well documented:

'But the yeere 1606, the fourth of King James, the ryver of Severn rose upon a sodeyn Tuesday mornyng the 20 of January beyng the full pryme day and hyghest tyde after the change of the moone by reason of a myghty strong western wynde,' wrote

John Paul, Vicar of Almondsbury. Stow in 1615 recorded that an inundation took place in the East Anglian Fens owing to a violent storm, and that further south in Romney Marsh, Kent the sea came in, 'so outrageously that it did not seem that the area could ever be reclaimed'.

The only authentic historical record of a big tsunami striking Britain was in 1755, when an earthquake off Portugal destroyed Lisbon. Tsunami waves reaching 10ft (3m) high hit the south coast of Cornwall, but with no known damage. And even this most extreme seismic event caused waves in southwest England no higher than those experienced during a typical storm surge.

But there was one catastrophic tsunami around 7,000 years ago, that swamped the Shetland Islands and Scotland's eastern coast. During survey work on a gas field in Norway's seas, the remains of a gigantic underwater landslide was found, known as the Storegga landslide. A huge chunk of land over 34,000sq miles (88,000sq km) under the sea had collapsed – about the size of mainland Scotland itself. That colossal slump may have been triggered by an earthquake, possibly detonated by an explosive release of methane from underwater gas deposits, creating a gigantic explosion that ruptured the continental shelf. Tsunamis of gargantuan proportions, up to 20m (70ft) high smashed into Scotland and drove inland for up to 50 miles (80km). The evidence for the disaster lay in a curious layer of sand, pebbles and deep sea diatoms extending along the northern and eastern coasts of Scotland. No storm surge has ever been known to produce anything like this and over such a large area. For Stone Age peoples that lived in the area at the time, the flood must have been catastrophic.

The disaster could happen again, though. There are many areas of potentially unstable deposits along the continental margin of northwest Europe, and at some locations surveys of the sea floor have picked out the traces of old submarine slides. Since 1983, three of the strongest earthquakes in northwest Europe have taken place within 62 miles (100km) of the Storegga slides, an area which is presently dormant.

Lorna Doone

Winter 1683

'It was a time when it was impossible to milk a cow for icicles, or for a man to shave some of his beard without blunting his razor on a hard grey ice.'

This was the terrible winter told by R.D. Blackmore in *Lorna Doone*, the story of a clan of outlaws, the Doones, who terrorise Exmoor. In this bleak landscape, farmer John Ridd falls in love with Lorna Doone, whom he rescues in a blizzard. 'And all the while from the smothering sky, more and more fiercely at every blast, came the pelting, pitiless arrows, winged with murky white, and pointed with the barbs of frost,' described Blackmore, himself a keen amateur meteorologist.

The story is based on the real winter of 1683/84, probably the coldest winter in British history, although temperature measurements in those days were crude and fairly unreliable.

In mid-December 1683, England was seized by an intense frost. The ground froze to 4ft (1.2m) deep, trees split open with frost and seaports in the English Channel froze over. 'It was a severe judgement on the Land: the trees not onely splitting as if lightning-strock, but Men & Catell perishing in divers places, and the very seas so locked up with yce, that no vessells could stirr out, or come in,' wrote the diarist John Evelyn. 'The fowls, fish, and birds, and all our exotic plants and greens, universally perishing. Many parks of deer were destroyed; and all sorts of fuel so dear, that there were great contributions to keep the poor alive,' he continued. The Thames in London froze down to

London Bridge, and possibly the greatest frost fair ever held lasted for over for two months, attracting thousands of people including Charles II. On 24 January John Evelyn noted: 'The Thames before London was still planted with booths in formal streets, and with all sorts of trades and shops furnished and full of commodities of all kinds, even to a printing press where the people and the ladies took a fancy to have their names engraved, and the day and year set down when it was printed on the Thames. Coaches plied to and fro as if in the streets, and there was sliding, bull-baiting, horse and coach races, puppet plays and other interludes, tippling and other lewd entertainments – so that it all seemed to be a bacchanalian triumph, or a carnival upon the water.'

But London also suffered an appalling smog problem. According to Evelyn, 'London, by reason of the excessive coldness of the air hindering the ascent of the smoke, was so filled with the fuliginous stream of the sea-coal, that hardly could any one see across the streets; and this filling of the lungs with the gross particles exceedingly obstructed the breath, so as one could scarcely breathe. There was no water to be had from the pipes or engines; nor could the brewers and divers other tradesmen work; and every moment was full of disastrous accidents.'

That bitter winter was probably caused by an anti-cyclone blocked over Scandinavia and sweeping down Arctic air, blocking out milder Atlantic winds. And its effects were probably far and wide. 'Nor was this severe weather much less intense in most parts of Europe, even as far as Spain in the most southern tracts,' reported Evelyn.

The 'Ill Years'

Famine of the 1690s

The weather took a marked downturn in the 1690s, with some truly extraordinary conditions. The evidence is littered through nearly all records across Europe – winters started early and turned horrendously bitter, springtimes were cold and late, summers were bad, wine harvests late and wheat prices rose with widespread reports of crop failures, disasters and abandonment of marginal farmland. Weather records in central England confirm the cold late springs in the 1690s, when diarist John Evelyn wrote of 18 May 1698, 'Extraordinary great snow and frost nipping and spoiling the corn.' It was the coldest May in England on record. The records of Dutch merchants show that the canals of the Netherlands froze solid for weeks and months on end and trade was stopped during the severe cold of the 1690s. Glaciers in Scandinavia grew larger and in 1695 Iceland was cut off from the outside world by sea ice for most of the year, devastating its farming and blockading its fishing fleets, leaving the population facing starvation. Conditions were so severe that the ruling Danish government made plans to evacuate the entire population of Iceland.

Across Europe, crops failed and marginal farmland on exposed ground was abandoned. Famine swept Europe. Following two bad harvests, France was struck in 1693 by one of the worst famines since the early 1300s. In Finland the springs were so cold that sowing could not be started before midsummer and then frosts began in late August before the crops could be harvested, and by 1697 famine in Finland killed around a third of the population.

Scotland was hit particularly hard in the highlands and eastern side of the country. It was called the 'Ill years of King William's reign', between 1693 and 1700. Travellers reported that the Cairngorm Mountains for the first time were permanently covered in snow. The seas were so cold that ice at one time stretched from Iceland to the Faroe Islands, just 200 miles (320km) north of the Shetlands, and cod disappeared from the seas. Eskimoes reached Aberdeen in kayaks, leaving the local people baffled with their incomprehensible language. They were called the Fin Folk, perhaps because they thought they came from Finland. 'His boat is made of seal skin, or some kind of leather; he also hath a coat of leather upon him, and he sitteth in the middle of his boat with a little oar in his hand fishing with his lines,' wrote the Reverend John Brand, a clergyman of the Church of Scotland. One kayak is still on display in University of Aberdeen Museum.

On 10 May 1698, an archive in Scotland recorded: 'An 'unkindly cold and winter-like spring' was threatening again to frustrate the hopes of the husbandman, 'and cut off man and beast by famine.' Already the dearth was greatly increased, and in many places 'great want both of food and seed' was experienced, while the sheep and cattle were 'dying in great numbers'. To the strict religious tone of the time, there was divine dimension to these hardships: 'In consideration of these facts, and of the abounding sins of profaneness, Sabbath-breaking, drunkenness, &c., whereby the displeasure of God was manifestly provoked,' a 'solemn humiliation and fast was ordered for the 17th of May within the synod of Lothian and Tweeddale, and the 25th day of the month for the rest of the kingdom.' It is no coincidence that this period in Scotland was marked by trials of witches, who were blamed regularly for bouts of poor weather throughout Europe during the 16th and 17th centuries.

In the Scottish uplands, the staple crop of oats failed and many people were reduced to eating wild mustard and nettles, and the lack of food led to typhus and dysentery epidemics. Death rates soared and birth rates fell, resulting in a loss of up to two thirds of the population, a worse rate of attrition than during the Black Death plague over three hundred years before. There are many descriptions of the horrors of these years – bodies lying by roads with grass in their mouths, or feeble souls dragging themselves towards the graveyards to be sure of a Christian burial. Sir Robert Sibbald, Geographer Royal for Scotland, in 1699 described the suffering he saw:

> 'Everyone may see Death in the face of the poor that abound everywhere; the thinness of their visage, their ghostly looks, their feebleness, their agues and their fluxes threaten them with sudden death if care be not taken of them. And it is not only common wandering beggars that are in this case, but many householders who lived well by their labour and their industry are now by want forced to abandon their dwellings.'

In Edinburgh, a refugee camp for the destitute was opened as the poor of Leith starved on the streets. Aberdeen was overwhelmed with death and disease, and soldiers were posted at the city gates to prevent famine refugees from outlying areas from entering and using up the city's meagre food rations. When this measure failed, a draconian edict was drafted to evict anyone who had not been resident in the city for at least seven years. In desperation, some 50,000 people across Scotland migrated to Ireland, particularly Ulster, where the climate was milder. Altogether, it took over 50 years for the population numbers of the country to fully recover.

The impact of the bad harvests was made worse by an economic depression, following wars, a serious downturn in trade with the Continent, high taxes, rapidly rising prices and political fallouts with England.

The 1690s may have suffered from cold water pushing down from the Arctic deeper into the North Atlantic, sending sea-ice around Iceland and towards the Faroes. This cold water and ice would have brought land temperatures down, and also steered Atlantic depressions and storms further south. Records of cod fisheries give another valuable insight into the severe cold: cod thrives in temperatures of between 4°C and 7°C (39–44.5°F) in winter but dies in temperatures below 2°C (35.5°F). In the Faroe Islands, no cod was seen for 30 years from 1675 onwards, and by 1695 it disappeared off the entire coast of Norway and became scarce even off the Shetlands.

The climate in the 1690s was also hit by violent volcanic eruptions in Iceland, the Pacific and Ecuador, that shot dust into the stratosphere and helped block out sunlight.

In desperation, the Scots sought political and economic salvation from an outlandish dream, the colonisation of the Darien jungle in Panama. They proposed setting up an international trading company there that would bring untold wealth to Scotland. After a huge public subscription and two fleets of colonists sent across to found the colony, the entire enterprise collapsed from disease, malnutrition and finally a Spanish counter-attack. The personal savings of thousands of Scots was lost at just the time that the famine reached its peak, and left Scotland in crisis. The only salvation was, in fact, the loss of Scottish independence in the Act of Union with England in 1707.

'The Great Frost'
Winter 1740

In 350 years of records, 1740 ranks as England's most severe and protracted winter. It was a shock to Britain and much of Europe, not only for its savage cold but also because it was so unexpected, after what had been a fairly benign climate that century, which had helped fuel a boom in farming and a rising population.

So when the cold arrived it came out of the blue. Ice floes bobbed around like small icebergs in the Straits of Dover, and stormy seas broke over ships and froze sailors to death.

Water wells froze, leaving a shortage of drinking water. Lakes and rivers turned to ice, and water mills could no longer work. Ships and boats could not break through frozen rivers, and so the price of coal rose eight-fold. So desperate were colliers (coal ships) to break out of the frozen Tyne that a channel had to be hacked by hundreds of men through the ice to open waters.

In Scotland, the *Caledonian Mercury* reported on 17 January: 'All the mills are at a stand, so that the price of meal etc. is risen as well as that of coals. They talk of people being froze to death on the roads.' By February the death rate in Edinburgh had more than doubled.

With transport by boat and road paralysed, bread became scarce and fetched such exorbitant prices that the poor were left starving. An eyewitness at the time observed: 'The fishermen, carpenters, bricklayers etc., with their tools and utensils in mourning, walked through the streets in large bodies, imploring relief for their own and families' necessities.' Landowner Robert Marsham wrote from his country home in the village of Stratton Strawless, Norfolk: 'This has ever since been called the hard winter: its severity occasioned scarcity, and that produced riots, which were not quelled in Norwich without military assistance and the loss of six or seven lives. ... With Ye high price of provision in general, the Weavers in Norwich, and some idle people in Ye Country, rose in riotous manner, and did some damage in Norwich and Ye neighbourhood, for which some were Hang'd at Ye next Assizes.'

Rioting over soaring food prices erupted in many other places, but in London the Lord Mayor of London launched an appeal to help the poor and even the parsimonious George II subscribed generously.

Thomas Barker, brother in law to the well-known naturalist Gilbert White, described how ice formed more than 3in (7.5cm) thick in only a day at his home in Rutland, trees were split open by frost, and vast numbers of small birds perished, which took years to recover their numbers. Landowner Robert Marsham in the village of Stratton Strawless, Norfolk noted rabbits starved in their warrens, beer turned to ice and at night his chamber pot 'froze to a cake'. When he tested the temperature outside by pouring water out of the window, it instantly turned to ice.

By February Windermere was so frozen that 'heavy-laden carts and droves of cattle' could cross the ice, and horse races were held on the frozen River Tees at Barnard Castle, County Durham.

A vicious gale in January drove ice on the Thames into a vast jumble that looked like 'icebergs rising on all sides in gigantic masses' and sank many boats and ships docked there. The river remained frozen in London for nearly seven weeks. It was a disaster for the port, as ships could not enter the Upper Pool, and those whose jobs depended on the port were left unemployed. Coal and many other essential goods became scarce, and the Lord Mayor was forced to launch an appeal to help the destitute from the big freeze.

Possibly the only good that came of the frozen Thames was a carnival on ice, a frost fair, with tented stalls selling food, coffee, alcohol and oxen roasted on open fires. There was good business for printers, who produced souvenir sheets with their customer's name printed with verse, much of it of dubious quality, such as this gem:

> 'Behold the liquid Thames now frozen o'er
> That lately Ships of mighty burden bore.
> Here you Print your name tho' cannot write
> 'Cause numbe'd with cold: 'This done with great delight.
> And lay it by; That Ages yet to come
> May see what Things upon the Ice were done.'

The frost fair was not all fun, though. Many Londoners fell over and broke bones or fell through thin ice and drowned. The

Icemen of The Royal Humane Society were established to deal with these tragedies.

The river remained largely frozen until 27 February, when the ice broke up into huge blocks that drifted off with all the booths and stalls of the frost fair and crashed into the old London Bridge, badly damaging its arches. But it remained so cold that it took the rest of the month for the ice to melt.

That savage winter was followed by a cold and very dry spring that brought even more problems in its wake. In March, disease and death seemed to afflict the country, and one doctor in Ripon, Yorkshire described skin eruptions among the poorer countryfolk that may have been scurvy. April continued very dry with hard frost at night that caused a severe water shortage for cattle and stopped water mills from working. The following month was one of the coldest Mays ever recorded in England, and Plymouth was hit by three days of snowfall. June was distinctly chilly and dry and wheat was so scarce that it sent prices soaring and led to food riots in Newcastle. Mobs seized granaries, and then stormed the crown-court and looted corporation money before troops quelled the rioters.

After the winter thawed out, Robert Marsham searched the countryside round his estate in Norfolk for the first signs of spring, but he did not hear a song thrush until well into March, and hawthorn blossomed two months late, in June. Small birds that were plentiful before the frost suffered terribly and it was three to four years before their numbers recovered.

The weather improved somewhat in the summer, although the diary of John Capps, of Biddesden, Wiltshire recorded: 'Wee had very little Grass & a very Poore Harvest... no Raine to Produce un till ye midle of Augst [when] ye Earth Produced a great Quantity more.' The harvest was late and more problems reappeared with one of the coldest Octobers on record in England.

The drought lasted from 1740 to 1743, one of the longest droughts in Britain, during which the death rate soared.

But conditions were far worse in Ireland, where a great frost led to a famine crisis almost as catastrophic as the infamous potato famine a century later, and became known as the forgotten famine. As in Britain, the cold killed vegetables, livestock, and froze up water-mills. When a thaw came in early February it was found that the extreme cold had destroyed the potato stores and there was hardly any to plant for the coming year. And like Britain, during the spring and early summer of 1740, Ireland experienced a parching, dry and bitterly cold easterly wind. Rainfall was only a fraction of what it ought to be, cattle fodder was almost non-existent, and the crops that had been sown withered in the drought. A series of strange seasons continued well into 1741. As famine tightened its grip, it became known as 'the year of slaughter', and is thought to have killed some half a million people, either from starvation or from dysentery and typhus that followed. The long drought eventually ended in September 1741 with a series of violent storms and exceptional floods, after which the weather returned more or less to normal. But the legacy of 1740–41 was a national catastrophe and waves of emigrants departed to North America. Apart from Norway, no other country in Europe suffered as much as Ireland from the drought and cold. In fact, that savage weather was the beginning of a downturn in climate that ended with a succession of bad weather and crop failures at the end of the 17th century.

Bantry Bay Storm

1796

The year 1796 was a terrible year for the British. Defences were fully stretched in the war with revolutionary France, leaving Ireland dangerously exposed – a convenient backdoor for a French invasion. In December that year a ferocious blizzard left people dead on the streets of Paris and London, but for Ulster rebel Wolfe Tone it was an ideal opportunity to catch the English off guard. So he persuaded the French to invade England by first landing in Ireland.

The week before Christmas, a French armada of 43 ships carrying some 15,000 soldiers slipped out of Brest past the British naval blockade and headed for the southern coast of Ireland. But rough seas separated the bulk of the fleet from their flagship carrying commander Lazare Hoche. The fleet waited on the outskirts of Bantry Bay for Hoche, which proved a fatal delay. A huge storm blew the entire fleet off anchor and out into the Atlantic, scuppering the entire invasion. Wolfe Tone escaped, but after another failed invasion attempt two years later he was captured and committed suicide.

Had the weather been better in 1796, the invasion could have easily overwhelmed the small English garrison in Bantry, and with only about 10,000 English troops in the whole of Ireland, the French would undoubtedly have swept through, supported by a sympathetic native population. That would have led the way to an invasion of Wales, leaving the English so vulnerable they would probably have had to strike a truce with the French. As a result, there would have been no Battle of

Waterloo, maybe no reunification of Germany and French may be the spoken language of Ireland and possibly even England today. As has happened so many times in history, the weather saved the English at just the right moment.

White Xmas

The classic picture of a snowy white Christmas is so untypical of what Christmas weather is really like – usually mild, grey and a bit of drizzle thrown in. So where did the passion for snow at Christmas come from? It is actually a nostalgic throwback to what winters used to be like, and we can thank Charles Dickens for that. In fact, you could say that Dickens is the man who invented the modern Christmas.

Born in 1812, Dickens would have experienced some very harsh weather in his childhood. Worst of all was the appalling 'Year Without a Summer' in 1816, when northern Europe was plagued with frost and snow even in June and July. The young Dickens would have seen snow on Christmas on six out of his first nine years, in what was the coldest decade in England since the 1690s.

Dickens worked the snow and frost into a whole series of stories and books featuring Christmas, including *Pickwick Papers*.

'Well, Sam,' said Mr. Pickwick, as that favoured servitor entered his bed-chamber, with his warm water, on the morning of Christmas Day, 'still frosty?' 'Water in the wash-hand basin's a mask o' ice, Sir,' responded Sam. 'Severe weather, Sam,' observed Mr. Pickwick.

Later that Christmas Day, members of the Pickwick Club ventured out on to the ice for skating at Dingley Dell. But was it based on any real weather?

Christmas in *Pickwick Papers* was supposed to be set in 1827, but the real Christmas that year was actually mild. In fact, the book was originally meant to depict Christmas 1830, which was dry, bright and freezing cold, with temperatures sinking to around −9°C (16°F). So, the icy weather at Dingley Dell was probably a fair reflection of the real Christmas.

A Christmas Carol was published in 1843, and again it is full of freezing cold weather. Small surprise, then, that Scrooge's Christmas Eve is singularly uninviting: 'It was cold, bleak, biting weather; the fog came pouring in at every chink and keyhole, and was so dense that, although the court was of the narrowest, the houses opposite were mere phantoms.'

> 'They stood in the city streets on Christmas morning, where (for the weather was severe) the people made a rough, but brisk and not unpleasant kind of music, in scraping the snow from the pavement in front of their dwellings, and from the tops of their houses: whence it was mad delight to the boys to see it come plumping down into the road below, and splitting into artificial little snow-storms.'

Dickens actually took a good deal of interest in the weather and accurately described it in his books. When he founded a newspaper called the *Daily News*, one of the features begun in 1846 was a weather column, the first of its kind anywhere in the world, written by a noted meteorologist, James Glaisher, at the Greenwich Observatory.

So, what made Christmas cold in those days? It was probably thanks to a spate of some severe volcanic eruptions during the early 1800s. Some of these shot dust high into the stratosphere, dimming the sun and cooling global temperatures. Most violent of all was the mega-eruption of Tambora in Indonesia in 1815, which pumped out huge amounts of ash and acid, and was responsible for the plunging global temperatures of 1815–16.

Once the smoke cleared the climate recovered, winters tended to grow milder. But the white Christmas bandwagon snowballed and the icy charms of *A Christmas Carol* struck a chord with the Victorians, nostalgic for the past and made the sentimental scenes on the first Christmas cards when they appeared in the 1840s.

These days, a white Christmas is rare, at least in England and Wales. Over the last century, London only had 11, so putting a bet on one is quite risky, even though you only need a single snowflake to fall on an official weather station. For many parts of the country you would be much better off gambling on a white Easter than a white Christmas.

Avalanche

1836

The worst snow disaster in British history happened in Lewes, Sussex over Christmas 1836.

Snow started falling over much of England on Christmas Eve and blew up into a furious blizzard on Christmas Day. Howling easterly winds drove the snow into a swirling white-out that blotted out almost every recognisable feature outdoors. The blizzard raged through into Boxing Day, piling up snow into monstrous drifts, roads throughout England were blocked for several days, and many lives were lost.

The snowfall was particularly heavy over the South Downs, and in Lewes snowdrifts reached up to 50ft (15m) high, cutting the town off completely.

A particularly menacing snowdrift built up at the top of a cliff and formed a hanging shelf of snow towering over a row of terraced houses below. When cracks appeared in the snow overhang on 27 December the residents below were told to evacuate their homes, but they ignored the warning thinking it was a joke. 'The poor creatures appeared to be bewildered and could not be prevailed upon to depart,' described the *Sussex Weekly Advertiser*. 'The snow toppled on the brink and sliding down the steep slope with tremendous force threw down and completely buried the seven end houses. The mass appeared to strike the houses first at the base, heaving them upwards, and then breaking over them like a gigantic wave to dash them bodily into the road.' Afterwards there was no sign of the dwellings. 'There was nothing but an enormous mound of pure white.' The snowshelf had avalanched some 350ft (100m) down onto the houses. Rescuers worked all day to dig through the snow and wreckage and found six survivors alive and in some cases completely uninjured. But they also discovered nine bodies that died either from suffocation or being crushed under the weight of snow and debris. You can still see where the disaster happened, at a pub called, appropriately, the Snowdrop Inn.

Tay Bridge Disaster

On 28 December 1879, a fearsome gale blasted through Scotland and caused the worst engineering disaster in British history, killing 75 people.

Winds gusted well over 80mph (130km/h), chimneys collapsed in Dundee and many areas flooded. The gale slammed into the Tay Bridge, connecting Dundee to the county of Fife, and whipped up water into swirling white spray. At 7.13pm a passenger train crossed the bridge, but midway across the wrought iron girders gave way, a half-mile section of the bridge collapsed, and the train plunged into the Firth of Tay. Astonished eyewitnesses described sparks showering out from the engine as it fell. The following day steam boats searched the waters for survivors, but only dead bodies were recovered. Everyone on board the train died, some 75 people, the worst train accident in British history.

The disaster sent shock waves through the public. The Bridge had been opened only for 18 months, and was seen as a triumph of Victorian engineering. It was the longest bridge in the world at the time, 2 miles (3.2km) across the Tay with 85 spans of 3,700 tons of cast iron held together by 2 million rivets and supported on plinths made from 10 million bricks. Afterwards a board of inquiry was hastily convened and concluded that the bridge failed through poor engineering and poor quality materials. The blame was pinned on Sir Thomas Bouch, the chief engineer of the bridge construction, for failing to allow for the effect of the wind on the design of the bridge, and for slack supervision. He was sacked in disgrace and died a few months later. To this day, though, there is still debate what caused the collapse and whether Bouch was to blame.

New evidence reveals that repeated warnings about a damaged girder were ignored in the months before the tragedy, causing trains to jump on the track, evidence that was never presented to the inquiry. In fact, the bridge was badly constructed after the builders went over-budget and cut corners to save time and money. Bouch was something of a scapegoat.

The disaster led to improved engineering standards and also more attention paid to measuring wind speeds and taking gale force storms into account in the design of large structures such as bridges. Much was made about how to measure the strength of the wind, and a new type of wind speed instrument was invented, the Dines Pressure Tube Anemometer, which became the standard wind-measuring instrument for decades afterwards.

London Flood

1928

When the Romans established Londinium they soon realised that the Thames was a dangerous river prone to flooding, so they built embankments along the riverbanks. In 1099 the *Anglo Saxon Chronicle* relates that 'On the festival of St Martin [11 November], the sea flood sprung up to such a height and did so much harm as no man remembered that it ever did before.' At Woolwich in 1237 the marshes were described as 'a sea wherein many were drowned' and in the Great Hall at Westminster Palace lawyers had to row around in wherries, row boats. In 1242 the river overflowed at Lambeth over 6 miles (9.5km), which would have included all the land up to and past Elephant and Castle, including, perhaps ironically, Waterloo.

Flood water receding from the Great Hall of Westminster Palace in 1579 left fishes gasping on its floor.

'There was last night the greatest tide that was ever was remembered in England to have been in this River, all Whitehall having been drowned,' wrote Samuel Pepys in his diary on 7 December 1663.

These days the threat of Thames floods is worsening. As Britain recovers from the last ice age, Southern England is sinking about 1ft (30cm) a century. On top of that, sea levels are rising because of global warming. As a result, tide levels are rising in the Thames Estuary relative to the land by about 23.5in (60cm) per century. Surge tides from storms down the North Sea are a particular threat: when a trough of low pressure moves across the Atlantic towards the British Isles, the sea beneath it rises above the normal level, creating a 'hump' of water that can sometimes threaten London.

This is how London came close to catastrophe in January 1928. Christmas had been unusually cold and snowy, and when the snows melted on 2 January it coincided with a terrific downpour of heavy rains, swelling the river. Then in a further blow, on 6 January a storm blew across Scotland, pushing water down into the funnel shape of the North Sea. As this storm surge pushed into the bottleneck of the English Channel it rose higher and pushed the swollen waters up the Thames Estuary, where they collided with the swollen waters rushing down the river. And in yet another freakish coincidence, the storm surge came at around the time of a high spring tide.

As a result, the Thames rose higher and higher until it burst through the river embankment walls. There was no warning in a dark, cold and windy night. Torrents of water rushed through roads, cascading down into basement flats, and drowned people in their beds. Those that did wake up in time fought for their lives, battling the surging waters and floating debris, imprisoned behind security bars on the outside of their windows. *The Times* told how one youth was drowned in a basement: 'the weight of water held his bedroom door fast, and the window was equally useless as an avenue of escape, since it was closely barred. Efforts to bend the bars and let him out were made by party of neighbours, and would have succeeded had the water risen less quickly. As their work neared completion the rescuers, working already in water that reached their necks, heard his cries for help die away, and soon the basement was quite full of the muddy waters.'

River embankments were collapsing at many places along the Thames. At 1am in Putney, a large block of flats by the river was inundated and two girls in a basement flat managed to escape through a small window and swim out in the swirling waters and scream for help. Neighbours in the flat above tied together sheets to rescue the rest of their family trapped in the flat and hauled them out.

The worst of the floods struck Millbank, near the Houses of Parliament. In those days it was a largely working class area with houses divided into flats crowded with families. A large stretch of river wall collapsed 'with a noise like an explosion, and the water, tearing up the paving blocks from the roadway, rushed across the road and deluged the houses on the opposite side,' reported *The Times*. 'In an incredibly short time the whole of this district was under water. Some warning was given by the police, who ran along the threatened streets hammering on front doors with their truncheons and ordering the occupants of basements into higher rooms or into the street, but even some of the police were taken by surprise as the flood surged up and then almost carried them off their feet.' Five people were drowned in the area, all trapped in basements.

By midnight flooding started in Battersea, Poplar and Greenwich and at Temple underground station. Sewage seeped out at Barking as the sewage works were overwhelmed. Warehouses, factories, hospitals, town halls were flooded. The Blackwall and Rotherhithe Tunnels were submerged and the moat around the Tower of London, normally dry, was filled with water. Water cascaded into the courtyard at the foot of Big Ben and the Old Palace Yard was left a foot under water. The ground floor of the Tate Gallery was flooded, destroying a collection of Landseer pictures, Turner watercolours and drawings. One worker at the gallery was trapped in the basement and policemen had to strip off and dive into the dark waters to rescue him.

Further upriver, the London Underground power station at Lots Road was under water, which put the Underground system out of action. Chelsea and Wandsworth gas works were partially flooded. There was even a fish caught in the police station in Battersea as it sank under water.

It was incredible that only 14 people drowned and some 4,000 people evacuated and left homeless, but London had come close to a catastrophic flood that could have killed thousands. The flood of 1928 was a stark warning that the capital was living on borrowed time, and something had to be done to defend against floods. An inquiry afterwards looked again at a proposal made in the 1870s for a barrage with locks to be built across the river estuary, from Tilbury to Gravesend. But the idea remained on the shelf, and the only improvements made after 1928 were raising the river walls higher. It would take another 60 years before London had a decent level of flood defence with the Thames Barrier at Woolwich, and it took another disastrous flood to shake the government out of complacency.

Icestorm

1940

Ice storms are rare in the UK, but the severest event in the record books is reckoned to be in January 1940, during the Second World War. The country had been gripped by an arctic freeze, the coldest winter for a century, with heavy snows, bitter winds, and to make matters worse food rationing was introduced on 8 January. The temperature fell to −23.3°C (−9.9°F) at Rhayader in Powys, Wales on 21 January – the lowest ever recorded in Wales. Then on 27 January a savage ice storm swept much of southern Britain, as freezing rain instantly turned to ice the moment it struck anything.

The freezing rain fell for two days and encased the landscape in ice, what looked like a world of glass. Trees looked like frozen waterfalls, weighed down until they shattered. 'Beech trees could be heard crashing down all night,' reported one eyewitness in Hampshire. 'The splintering of the ice-casing made even more noise than the rending of the wood, like broken glass.' The iced leaves of evergreen shrubs made a noise like castanets as they rattled in the wind.

Dorothy Seton-Smith in Cheltenham remembered: 'Birds froze on the boughs of the trees, which looked peculiar, and hens in henhouses froze on their perch.' Frozen pheasants and rabbits could be caught by hand. Windows and doors of buildings froze solid, and some car drivers found themselves trapped inside their frozen vehicles. Umbrellas were useless as the freezing rain instantly turned solid with ice. Virginia Woolf wrote in her diary: 'The grass is brittle, all the twigs are cased in clear, brown cases and look thick, but slippery, crystallised as if they were twigs of fruit as dessert.' And she continued, 'Unable to go to London... On Sunday no cars could move... trains hours late or lost... almost out of meat.'

Thick coats of ice brought down power, phone and telegraph lines.

The Thames froze over at Kingston, Surrey for the first time since 1880, ice covered stretches of the Humber and the sea froze at Bognor Regis, Dungeness and Folkestone.

Troops leaving for France in the British Expeditionary Force faced huge problems simply keeping their vehicles running as anti-freeze froze in radiators. The sub-zero temperatures were highly dangerous for manufacturing explosives, and on 18 January 1940 a huge explosion ripped through the Royal Gunpowder Factory at Waltham Abbey, Essex, blowing the roof off the local Abbey church and wrecking shop fronts in the town, in an explosion that was heard as far away as Brighton. Five workers at the factory were killed.

But news of the scale of the ice storm was restricted by censors until much later to prevent the enemy learning about it. However, most of Europe was in the grip of an even more savage winter that brought much of the war to a standstill. This bought valuable time to reinforce defences in Britain, and stormy weather out in the Atlantic also brought some relief to Allied merchant shipping as German U-boats found conditions too difficult to operate. The main area of conflict in the war was the Russian invasion of Finland in unbelievably cold temperatures, below −35°C (−31°F). Outnumbered by some 40 to 1, the Finns used the snow and bitter cold to fight a guerrilla war that inflicted huge casualties. The Russians are thought to have lost some half a million men in a three-month war that they eventually won. However, it taught them valuable lessons in re-equipping for winter warfare.

Winter

1947

After the Second World War, Britain was bombed out, bankrupt and exhausted. Desperately short of fuel, the winter of 1947 proved a turning point that sank the country to a new level of deprivation unknown even during the war. As weather records go, it was hard to beat: the snowiest winter in living memory, the dullest winter, the coldest February, and the wettest March on record, ending in some of the most devastating floods of the century. This catalogue of weather calamities precipitated a national crisis and changed the face of Britain as well as the rest of Europe, for decades afterwards.

The winter began deceptively, with just a brief cold snap before Christmas 1946 and again in the New Year. Snow lay thick on the ground when in mid-January temperatures soared so high it felt like spring had arrived early. Unfortunately the snow thawed so rapidly it set off floods, just as a gale with hurricane-force winds brought down roofs, trees, and even collapsed houses and a railway bridge in Birmingham. Much of the Fens turned into a vast inland sea as frozen lumps of snow and ice blocked hundreds of drains and dykes, and a million people in East London had their water supplies cut off by a flooded water pumping station. Windsor was marooned in floodwaters and the town's gas works were put out of action, and the *Daily Telegraph* described a 'second Dunkirk' as ancient motor boats, river craft, motorised punts, dinghies and amphibious army trucks and jeeps were recruited for evacuations or to deliver fresh supplies of food and water.

But the thaw was a mirage. The real winter arrived shortly afterwards as the country was gripped in an Arctic freeze that lasted two months, with snow whipped up into monstrous snowdrifts that buried roads and railways. On 4 February *The Times* reported that a snowstorm buried hundreds of vehicles in Lincolnshire, including three buses that completely disappeared under snow, leaving the passengers inside stranded all night. Trains were lost in snow, and even relief engines sent to their rescue became stuck in snowdrifts. The temperature sank to –21°C (–5.8°F) at Woburn, Buckinghamshire. On 20 February the Dover to Ostend ferry service was suspended because of pack-ice off the Belgian coast. It became the coldest February ever recorded, and, to add to the misery, there was virtually no sunshine for almost the whole month. This was probably not the best time for the Government to continue with its economic programme of nationalisation, the public ownership of the electricity industry. A furious row in Parliament forced the Government to debate the fuel crisis, which Antony Eden, opposition speaker, described as the gravest industrial crisis the country had faced since the General Strike.

The freeze paralysed coal mines, the coal stocks were often stuck at the collieries by railways and roads buried in snow. Even carrying coal by sea was plagued by iced-over harbours and storms and fog at sea. At one stage more than 120 ships with 195,000 tons of coal were stuck at the vital coal ports of Tyne and Wear.

As stocks of coal rapidly dwindled, the nation faced a crisis unprecedented even during the war. The nation's heat and power almost entirely depended on coal – power stations and gas works were fuelled by coal, industry largely ran on coal, and railway engines needed coal. Homes were heated by coal fires and electric heaters, and were so badly insulated that much energy was lost, and many households could only afford to keep one room warm.

Just a week after the freeze began, the Minister of Fuel & Power, Emmanuel Shinwell, ordered electricity supplies to be cut to industry, and domestic electricity supplies turned off for five hours each day, to conserve coal stocks. The country was plunged into darkness and even Whitehall and Buckingham Palace were reduced to working by candlelight. When power was available it was often at a reduced voltage that only gave dim light. Gas pressure to homes was reduced so low that consumers were warned to watch out for weak flames that could easily blow out and cause gas poisoning or explosions. Traffic lights were dimmed or went out, many schools were closed and only essential buildings such as hospitals were allowed a full allowance of fuel and electricity. The television was closed down, radio output was reduced, newspapers cut down in size and magazines such as *The Economist* and *Spectator* ordered to stop publishing. The emergency package, however, hardly made much difference to power supplies but was a crushing blow to public morale. 'We could be worse, but we should be a lot better considering we won the war,' wrote a Wembley housewife, Rose Uttin, in the Mass Observation project.

People queued up at coal depots or gasworks with prams, buckets, sleighs and anything else they could carry hoping to get some meagre supplies of coal. One woman in Devon thought she had struck lucky when she spotted a stray bag of coal sticking up through the snow. It was the top of a coal lorry almost completely buried in snow. A musical instrument shop in Bristol offered pianos damaged in the war for firewood.

To add to the crisis, food supplies shrank to alarming proportions and rations were cut even lower than during the war. Farms were frozen or snowed under, and vegetables were in such short supply that pneumatic drills were used to dig up parsnips from frozen fields. For the first time ever, potatoes were rationed, after some 70,000 tons were destroyed by the cold. Even during the war they had not been rationed, and now the allowance was 3lb a week, later cut to 2lb. Bread rationing, only introduced in 1946, grew worse, and loaves were filled out with barley to eke out the miserable supplies of wheat. Some four million sheep and lambs died, almost a quarter of the national total, and 50,000 cattle died. Butchers found it difficult to cut meat that was frozen hard, even though supplies were so scarce that each person was rationed to just one shilling's worth each week. Fishing fleets that could get to sea found their nets and catches frozen solid when they were hauled aboard. The government tried a deeply unpopular campaign to encourage everyone to eat a cheap South African fish called snoek, millions of tins of which had been imported, but it tasted disgusting and was used eventually as cat food. Food supplies were in any case battling to get through blizzards and snow-drifts. The Attlee government seriously worried that the country could slide into famine. As Winston Churchill, then the leader of the Opposition, announced on 12 March: 'Let us put a fight for John Bull's food,' he said, 'To run him down as low as this is a scandal.' The deprivations lasted long after the winter, and at Christmas that year puddings were made with dried eggs and carrots, served with margarine and synthetic rum.

The nation remained frozen for seven weeks, but it was not unremitting cold everywhere. Tantalising glimpses of warm weather occasionally set off thaws, although these caused further problems as frozen water pipes burst causing serious water shortages, and gas pipes fractured killing several people with fumes.

The warm spells were only short-lived. Incredibly, March turned out even worse than February, made worse by appalling weather forecasters that predicted the end of the freeze. Instead, 5 March brought the worst blizzard of the entire 20th century, which raged for 48 hours with colossal snowfalls in hurricane-force winds. Snowdrifts up to 9m (30ft) high paralysed virtually the entire country and snowploughs, troops, road and rail workers and German prisoners of war struggled to clear the snow. Supplies of food grew so critical that in some places the police asked for authority to break open stranded lorries carrying food cargoes. *The Times* on 6 March reported 'The blizzard has virtually cut England in two. It is almost impossible to get from south to north.' A freak ice storm encased a large part of southern England in ice so thick it brought down phone and power lines, broke trees and reduced roads to ice rinks.

Eventually on 10 March a sustained thaw set in, and triggered another spectacular disaster. After weeks of deep frost, the ground was so hard that the melting snow ran off into raging torrents of floodwaters. And to make things worse, a huge storm dropped torrents of rain. Indeed, it was the wettest March on record in England and Wales.

The winds whipped up floodwaters into waves that breached dykes in the Fens, flooding 100 square miles (260sq km) of rich farmland, and houses collapsed under the torrents of floodwater. Canada sent food parcels to stricken villages in Suffolk; the prime minister of Ontario even offered to help dish them out. The banks of the Trent burst at Nottingham, submerging miles of streets and hundreds of homes. Army amphibious landing craft and an assortment of boats were used to evacuate people marooned in houses. A reporter in the *Nottingham Guardian* wrote 'Knee deep in swirling water, road after road was silent and deserted, the lights from scores of upstairs windows reflecting in the floods below'. When the floodwaters reached the tidal part of the River Trent, they hit a spring tide, and inundated the whole of the lower Trent valley. Bakers in Upton upon Severn and Shrewsbury delivered bread by boat, throwing loaves to upstairs windows, and the Australian Red Cross came to the rescue in Gloucester. In Chiswick, West London, small boys ran a thriving business shopping for marooned neighbours in boats made from zinc bathtubs. The Thames at Sonning, Berkshire burst its banks and fanned out 3 miles (5km) wide. Across the country, more than 100,000 buildings were damaged, over 1,000 square miles (2,600sq km) of land flooded across 40 counties and the damage was estimated at between £3bn and £4.5bn in today's prices.

It is difficult to imagine a worse run of weather, although the Government was blamed for the food and fuel crises. Elected in the summer of 1945 with a landslide majority, the Labour administration had embarked on a radical programme of nationalisation, including the heath service, coal mining, electricity supply and railways. But they were caught unprepared when people began to buy electric fires and immersion heaters and power stations could not meet the rising demand for electricity. Long before winter set in, stocks of coal were dangerously low and although the government was warned of the dangers ahead, Emanuel Shinwell brushed them aside, and in the previous October he announced: 'Every one knows that there is going to be a serious crisis in the coal industry, except the Minister of Fuel and Power. I want to tell you there is not going to be a crisis in coal, if, by crisis, you mean that industrial organisation is going to be seriously dislocated and that hundreds of factories are going to be closed down.' His faith was based on a hopeless pipe dream: a mild winter and coal mines working at peak capacity. For the beginning of December it seemed like the weather and power supplies might hold out. But when the big freeze set in, the government was left hopelessly unprepared. By a strange quirk of fate, though, the cold weather coincided with the nationalisation of the coal mines on 1 January, and the power cuts were seen as gross incompetence.

Even worse damage was done to the Government's credibility by the effect of the power shortages on the national economy. At the end of the war, Britain was the world's biggest debtor nation, owing about £3 billion and the balance of payment deficit – the difference between exports and imports – soared to £750 million. As industry closed down, around 2 million people were put out of work and the economy lost £200 million in desperately needed exports.

Yet despite the collapsing economy and the threat of starvation, the government carried on behaving as if it was a world superpower. Military expenditure was 15 per cent of GDP, far higher than before the war, and included the development of Britain's own nuclear bomb and forces stationed in Europe and across the Empire. With a hugely ambitious programme of free healthcare and reconstruction, it was simply unsustainable. The winter of 1947 led to savage cuts in public spending at home, and contributed to the humiliating devaluation of sterling from $4 to $2.80 the following year.

Farming suffered long-term damage, too. It took several years for the numbers of livestock to recover from the casualties of the 1947 winter, although some upland farms were so heavily decimated that they gave up sheep farming and sold their land to the Forestry Commission.

The winter of 1947 was a turning point in politics. Less than 2 years since winning the war, the nation was left freezing cold, plunged into darkness and on the brink of starvation, and for many people it showed that national planning and socialism did not work. It was the end of the honeymoon for the new Labour government, and some historians say it sounded their death knell, and Labour was turned out of office in a landslide defeat at the next general election.

But had the winter of 1947 been much kinder, and power-cuts avoided, perhaps Labour and its programme of nationalisation would have been seen as a great success. Perhaps it may have been seen as the natural party of government, as happened for the following decades in Norway, Sweden and Denmark.

The legacy of that winter lasted in other ways. After years of war and deprivation, the British public had had enough. Many simply voted with their feet and emigrated rather than suffer any more hardships. As one couple of emigrants to Australia commented years afterwards: 'We lived through the war, the shortages and then we jolly well nearly froze to death in 1947. When we were offered the chance to go somewhere warmer we couldn't say yes quick enough'.

Some historians believe that the winter of 1947 was also a milestone in the decline of Britain as a world superpower. The nation could hardly feed its own population let alone the starving millions it was responsible for in Germany, where the winter was even more savage. The populations of the bombed-out cities were reduced to an almost stone age existence of scavenging for food and fuel to survive. The Americans were horrified when they saw pictures of western Germany on the verge of collapse, and feared the country was sliding into chaos, ripe for a communist takeover. Together with a surge in communist parties in France and itay, the USA looked to Britain as the bulwark against the threat of communism in Europe. But instead they saw a nation on its knees, and grew even more alarmed when the UK withdrew support for anti-communist regimes in Greece and Turkey. It was then that the US administration realised they would have to save Europe single-handedly. As a result, the US proposed a more active role in the defence of western Europe and used the Marshall Plan to boost the recovery of the European economies with billions of dollars of aid. It was the Marshall Plan that kickstarted the Germany into its great postwar industrial revival and brought western Europe together into economic co-operation, which eventually grew into the Common Market.

The Great Smog

London 1952

London was always famous for its fogs, but in December 1952 it was struck by something completely out of the ordinary. A thick fog and clouds of coal smoke combined into a toxic cloud of acid and soot that paralysed the capital and killed thousands of people. It was called the Great Smog, the worst episode of air pollution ever recorded in the world, and a turning point in the history of public health that changed attitudes to the way that the environment was treated.

It began on Friday, 5 December 1952, although strangely it started off as a gloriously sunny day, crisp and cold with hardly a breeze in the air, and a huge relief after weeks of bad weather. But the cold led to millions of people stoking up coal fires to stay warm, and the coal smoke added to the coal being burnt from factories, power stations, railways and many other chimneys. Gradually, a thin brown haze built up over the sky, although nothing exceptional for winter in the capital.

As the sun set that evening, bitterly cold air sank to the ground, which was wet after weeks of rain, turning the moisture into fog that rapidly soaked up the coal smoke to form a thick mass of black smog.

Lights across the capital were blotted out, and as pubs and cinemas emptied later that night the crowds found themselves lost in a frightening world, where familiar streets had vanished. Traffic ground to a crawl as the blanket of smog grew thicker and more claustrophobic. 'The smog hit us like a wall,' remembered Barbara Fewster, who tried to get home by car across London that night. 'It was a terrifying journey. The only thing to do was for me to walk in front of the car and guide my fiancé, who was driving, and who was hanging out of the window.'

The smog was amazingly shallow, though. As cold, still air close to the ground filled up with smog, an invisible and stubborn layer of warmer air sat around 650ft (200m) above. This acted like a lid on a box, trapping the stagnant air below, a situation known as a temperature inversion. The winter sun was too weak and the wind too calm to break this stranglehold of warm air and blow the smog away. 'The total stagnation of the cold air sitting close to the ground simply couldn't rise and the air pollution stayed trapped in this very thin layer close to the ground,' explained Peter Brimblecombe, an air pollution expert at the University of East Anglia. In fact, the smog was so shallow that people walking across Hampstead Heath on high ground in north London could see the smog below, looking like a blanket of black cotton wool, whilst above it the air was so crystal clear they could see all way across to the hills of Surrey.

Of course, London smogs were nothing new, but this one was unique. A persistent block of high pressure centred over southern England becalmed the weather with hardly a breeze in the air. This allowed the temperature inversion to trap cold air and smog on the ground. In fact, it was so cold that temperatures remained at or slightly below freezing in central London and actually gave heavy frosts at night.

And because it was so cold on the ground, more and coal was burnt, pumping out even more coal smoke to add to the toxic stew brewing up. The day after the smog first appeared there was no sign of daylight, only blackness. Roads became littered with abandoned cars or crashed into each other, trees or streetlamps. Heathrow Airport closed down in visibility of less than 10 yards. Trains, road and river transport all stopped. 'We got on a bus but the journey was very hazardous – the conductor walked in front of the bus but we ended up in a side turning by mistake and nearly turned over,' recalled Betty Crowhurst in south London.

'The smog was so bad that you couldn't see your hand in front of your face. Drivers could see even less, all they could do was follow the lights of the vehicle in front. One of our dray lorries, with the drayman guiding it through the smog, turned into the brewery – unfortunately all the traffic in Commercial Road, including double-decker buses, followed it into the yard. The following morning it took hours to free the deadlock,' recalled Peter Prentice, in the East End of London.

At Sadler's Wells, the opera La Traviata had to be abandoned when the theatre was filled with smog, and cinemas closed when audiences could not see the screens. Horse racing, football matches and just about every outdoor sporting fixture was called off.

Many people were reduced to groping their way blind through the suffocating gloom. 'I lost myself in a street in London which I knew like the back of my hand. I couldn't see anything, had no idea where I was, and had to go to a wall and feel my way, until I could find a sign giving the name of the street,' remembered Sir Donald Acheson, formerly Chief Medical Officer.

Prize livestock at the annual Smithfield Cattle Show at Earls Exhibition centre collapsed dead. Interestingly, animal on dirty bedding survived, possibly because the ammonia in the straw neutralised the acid in the air.

Yet despite the horrific conditions, people seemed to accept the foul atmosphere. After all, London had a long history of smog. In 1272, King Edward I tried to ban the burning of coal in the city because of the foul air but his prohibition failed to make any difference, and so did attempts at smoke control by Richard III, Henry V and Elizabeth I. In 1661, the diarist John Evelyn complained bitterly about a 'hellish and dismall cloud of sea-coal' over the capital. 'Inhabitants breathe nothing but an impure and thick Mist accompanied with a fuliginous and filthy vapour.' By the 1800s, the smogs were so foul they became known as peasoupers for their greeny, yellow-brown colours. But the peasouper smogs also inspired the famous surreal paintings of London by Turner, Whistler and Monet. Authors such as Charles Dickens called it a 'London particular', as he wrote in *Bleak House* in 1852: "'I asked him whether there was a great fire anywhere? For the streets were so full of a dense brown smoke that scarcely anything was to be seen."..."Oh dear no miss," he said, "This is a London particular.'"

But the smog of 1952 was different from previous episodes. Instead of lasting a day or two, it persisted almost a week and grew into a toxic cloud of terrifying proportions: 1,000 tonnes of smoke particles, 2,000 tonnes or carbon dioxide, 140 tonnes of hydrochloric acid, 14 tonnes of fluorine compounds, 370 tonnes of sulphur dioxide.

In five days it became the worst single incident of air pollution recorded anywhere in the world. It was a man-made disaster borne out of a government order that high quality coal should be exported, to earn desperately needed foreign currency. The only supplies left for domestic use were cheap, low-grade coals full of sulphur, which burned poorly and gave off clouds of sulphur dioxide, which reacted with the fog to form sulphuric acid. In fact, the smog had the same acidity as a car battery.

When the fine droplets of acid and soot were breathed in, they inflamed bronchial tubes and lungs, causing serious bronchitis, emphysema and pneumonia. Many people collapsed dead at home, literally choked to death by the smog. Ambulances and doctors could not reach the victims in time because the roads were impassable, and those victims fortunate enough get to hospitals rapidly filled up the wards.

Only later did it become apparent that the number of deaths during the smog was far above normal, mostly from respiratory or cardiac causes. But the death rate soon rose dramatically and estimates afterwards put the total at 4,000 people killed. Morticians ran out of space for the dead bodies, undertakers ran out of coffins and florists ran out of funeral bouquets. Coroners, pathologists and registrars of deaths all laboured under huge workloads. In fact, the number of deaths soared to levels not seen since the devastating influenza epidemic of 1918.

On 9 December, something incredible happened, though. The smog suddenly vanished. A strong wind blew the lid off the temperature inversion and swept the whole filthy mess of toxic air out to sea. The smog had lasted continuously for 4 days and 18 hours, a new world record duration for smog caused by coal smoke.

However, a recent re-assessment of the death statistics from the smog revealed it was much worse than previously thought. Long after the smog, its aftermath persisted as the death rate remained abnormally high through the rest of that winter, spring and even into summer. People had developed complications that only later proved fatal. In fact, the final death toll caused by the smog is now estimated at over 12,000.

Another misconception at the time was that the smog victims were only the old and sick. Although two-thirds of the victims were over 65, the death rate among 45–64-year-olds also rose three times higher than normal. And there was also an alarming number of infant deaths.

The Great Smog of 1952 was a wake-up call to clean up coal smoke, yet the attitude of the government was to deny that it had any responsibility. Local government minister Harold Macmillan, later to be Prime Minister, complained: 'Today everybody expects the government to solve ever problem. It is a symptom of the welfare state. For some reason or another "smog" has captured the imagination of the press and people... I would suggest we form a committee. We cannot do very much, but we can seem to be very busy.' Ian Macleod, the Minister for Health, protested: 'Really you know, anyone would think fog had only started in London since I became a Minister.'

A recent study of government papers has revealed that the politicians even tried to blame the extra deaths on a flu epidemic, although there was no epidemic. Ministers decided that in any future smog crisis, doctors should give up to 2 million gauze smog masks to people with heart and respiratory problems, although health minister Iain Macleod admitted that the idea was 'only a gesture' and no known mask could truly protect from smog.

Despite the goverment's intransigence, a groundswell of public opinion and unease amongst many MPs forced the issue. A private member's bill in parliament to clean up air pollution jolted the government into announcing an inquiry, although they hoped it would sweep the problem under the carpet. The strategy backfired, though, when the committee called for the total ban on coal fires, and reluctantly the government brought in the Clean Air Act in 1956, authorising local authorities to force smokeless coal burning in their areas. In any case, the age of coal was declining as people switched to gas, oil or electricity for their heating, railway steam engines were scrapped for diesel or electric trains and industry largely gave up burning coal.

The results of banning coal smoke were startling. Visibility improved and the hours of daily sunshine increased. In the early part of the 20th century, central London averaged just 38 hours of sunshine in November; by the end of the century that had almost doubled to over 70 hours' sunshine. Vegetation such as lichens, which are particularly sensitive to air pollution, re-colonised the capital. Buildings were scrubbed clean of layers of black soot and revealed sparkling colours not seen in ages.

The Clean Air Act was a milestone in legislating against pollution, and was possibly the beginning of the modern environmental movement. But although the coal smoke has gone, another form of smog is choking cities, from traffic pollution and burning other fossil fuels, causing respiratory and heart problems in many people. Researchers estimate that modern air pollution causes more than 24,000 deaths each year in Britain. The lessons of the Great Smog of 1952 have not been completely learned yet.

Flood

1953

At 1am on 1 February 1953, in bitter cold, pitch black darkness and howling winds, prefabricated bungalows in Felixstowe began to float down roads like doll's houses, people clinging onto their roofs for dear life. All along the east coast of England the sea was invading in a horrific storm surge, the worst natural disaster of the 20th century in Britain. The floods killed 307 people and 32,000 were made homeless.

Earlier in the night, the storm tore over the Irish Sea and sank the ferry *Princess Victoria* bound from Stranraer in Scotland to Larne in Northern Ireland. The ship tried to turn back into port in the wild seas, but the monstrous waves overturned her with the loss of 133 lives.

An Atlantic storm tore across northwest Scotland growing more intense as it headed towards Denmark. On Orkney winds gusted up to 126mph (200km/h), and thousands of trees were felled across Scotland. The combination of winds and severe low pressure piled up a vast bulge of seawater down the North Sea riding on a high spring tide. As this storm surge was funnelled down the North Sea, it became squeezed into the shallow bottleneck between East Anglia and the Netherlands and seas rose up to 10ft (3m) above the predicted tides. At King's Lynn, Norfolk, the sea level was 7ft (2.1m) higher than a normal high tide and rushed through the centre of the town.

It all came without warning. Many telephone lines were brought down by the winds, so warnings about the storm could not be passed down from the northerly parts of the coast. And there was no single authority responsible for flood warnings, so co-ordination was shambolic.

In the early hours of the morning, the pressure of the rising storm surge, high tide and crashing waves was too much for the hundreds of sea defences stretching from Yorkshire to the Thames Estuary. After years of neglect, the crumbling earth barriers and solid walls simply burst under the onslaught and some 330 square miles (850 sq km) of land was flooded. Many people were drowned in their beds, others died of hypothermia as they clung on to the roofs of their flooded homes. For days afterwards, rescue workers collected bodies.

'I could hear another noise above all this, above the wind, and I pulled the curtain aside and looked out and everywhere was bright moonlight glistening like glass, all the way round me,' described Malcolm MacGregor of Lee-over-Sands, Essex. 'I couldn't make out what was wrong. The glass effect was the water running down the sea wall.' The sea wall 2 miles (3.2km) away had burst and the land behind was filling up with water. 'My father put my mother and brother in the boat and my sister and it was bitterly cold. Away we went and on the way across the marsh we gathered up our little pony and that little old thing swam behind all the way to the mainland. All the time we were doing that there were telegraph poles, haystacks, all sorts of things roaring past us.'

Colin Turner was a 19-year-old national serviceman at RAF Swanton Morley, Dereham, Norfolk when an emergency call for help came. 'Eight of us were dropped in Heacham and told to do what we could. It was 4am, pitch dark, you could hear the waves crashing. It was terrifying and bitterly cold, with the wind blowing a gale.'

'We could hear people crying for help. We tied ropes round our waists and waded in. Some were in upstairs rooms, balanced on the furniture, some were on the roofs of bungalows. I have no idea how many we pulled out.'

'The sea just came up and caught them. We plunged into the water after them. We were up to our necks. Soaked, but we just kept on all night.'

'In the morning we waded out to see if there was anyone left alive and we could see the drowned people floating in the bungalows.'

Richard Lord was seven years old, and living in Felixstowe. 'You could hear people screaming, things breaking. I do remember a cow going past the window, whether it was dead or alive I don't know.'

> 'My stepfather said the only thing we could do was climb out of the bedroom window and onto the roof. So my two parents were obviously very wet by now, up to their waists, and I was held while my stepfather opened the window. Mother pulled me up onto the roof and then he followed on behind. We lay on the roof, my parents either side of me to keep me warm. You could still hear obviously screaming, shouting. The water rushing by, trees and bits and pieces. The prefab then lifted up off its base and started to move and as it started to move we moved forward several blocks. We just sat up there all night.'

London came within a hair's breadth of flooding. As the storm surge raced up the Thames Estuary it burst into docklands, from Tilbury to London's East End, flooding oil refineries, factories, cement works, gasworks and power stations. In the East End, large sections of river wall collapsed, flooding more than 1,000 houses in West Ham. But the floods further out in the Estuary and along the east coast had eased the pressure in the storm surge just enough to save London from flooding. It was a desperately close escape.

The cost of the damage is estimated at £5 billion in today's prices, and sent a shockwave through the country. As with the London flood of 1928, the government immediately set up a committee to examine the risk to the capital from flooding. To begin with, the loss of life was so great that it was recognised that something had to be done, and recommendations were made to build a barrier across the Thames to stop storm surges from the North Sea invading the capital. But from then the bureaucracy crawled at a painfully slow pace, and the politicians dragged their feet. Several designs of flood barrier and locations were examined, but they were all hugely expensive and decisions were delayed. As the years went by, though, public unease grew steadily, and when the Greater London Council was created in 1965 it took over responsibility for flooding and added impetus for the barrier project. Eventually the go-ahead for the Thames Barrier at Woolwich was given in 1970 and it was opened in 1982.

However, the Dutch learned the lesson from 1953. The floods swamped about a tenth of their land surface, left half a million refugees and killed some 2,000 people. The nation woke up to the fact that they were living on borrowed time, and built the world's biggest flood defences. Even now a fortune is spent each year maintaining the flood defences.

In Britain, flooding is not such a pressing issue, yet even this flood defence is a temporary measure. Southeast England continues to slowly tilt downwards 1ft (30cm) every 100 years, and sea levels are rising as glacier and ice sheets melt, and as seas grow warmer they expand. The Barrier was designed to last until 2030 and then undergo a refit to raise its flood gates higher, but sea levels may be rising faster than anticipated. In 2007, assessments began on a new flood defence project, but it remains to be seen what will come of these plans. If it is anything like the development of the Thames Barrier it will take another 30 years before anything is done.

London is one of the most vulnerable cities in the world to flooding. A huge amount of infrastructure is vulnerable to inundation: 38 underground stations, eight power stations, 16 hospitals, 400 schools, over 30 phone exchanges, Belmarsh maximum security prison plus half a million other buildings. The Thames Gateway project with tens of thousands of new homes in the floodplain of the Thames raises the risks even higher. An estimated £125 billion worth of properties are vulnerable, added to which are the costs of disruption to the economy, telecommunications and government. The knock-on effect could plunge the national economy into recession. More than 1.25 million people live or work in the Thames floodplain, and if they all had to be evacuated it is anyone's guess how the transport system would cope.

The Thames Barrier was designed to withstand a one-in-1,000 year extreme storm surge event, but the Dutch engineered their defences ten times greater. Sea levels are rising rapidly, and a new study reveals that London is sinking even faster than previously thought, whilst sea levels could be rising faster. The trouble is that the Barrier has been so good at protecting London that it has rather been taken for granted. It is due for an upgrade in 2030, but the clock is ticking away before a new flood defence is needed. The Environment Agency is considering whether to build another hugely expensive barrier, or allow coastal flooding to relieve the pressure on the Thames. And not only the capital needs protecting, because coastal defences also need strengthening around towns and cities such as Grimsby, Hull, Lowestoft, Southend, Great Yarmouth,

Hastings, Portsmouth and Cardiff. If the bureaucracy and foot-dragging after 1953 is anything to go by, it will be another 30 year before the next flood defence is ready for action.

Sheffield Storm

February 1962

In the early hours of 16 February 1962, a fierce Atlantic gale raced across the Pennines and tore down into Sheffield with astonishing and devastating hurricane-force winds. 'Sheffield this morning is a city in chaos, numbed and paralysed by the horror of a 96mph hurricane,' reported the *Sheffield Telegraph*. 'It's just like the Blitz.' Around 100,000 homes, two-thirds of the city's housing stock, were damaged, 100 beyond repair. Roofs were ripped off, chimneys crashed down and windows smashed. One family were left sitting in their living room with only the front wall standing. Even a new semi-detached house was almost demolished. 'We are completely open here and the wind was making the entire building shake. There was a terrific gust and the roof just disappeared,' one man told the *Sheffield Telegraph*.

A 130ft (40m) tall crane crashed onto a building site at the Technology College and one of the floodlight pylons crashed down into a stand at Sheffield United's football ground at Bramall Lane. Buses and trucks were flipped over and roads were blocked with bricks, slate, glass and fallen trees. Three people were killed by falling masonry.

The Government declared the city a disaster area and schools and church halls opened as emergency accommodation for homeless families.

As the gale blew over the Pennines the winds had become squeezed as they were trapped between a layer of warm air on top of colder air – a temperature inversion. The winds bounced up and down in this atmospheric sandwich, and, rather like air squeezed in a bicycle pump, the air became compressed into a jet of fast wind.

Winter

1962–63

The winter of 1962–63 was the coldest since 1740, and only missed out beating that winter by a whisker. So ferocious was the 1962–63 winter that there was serious thought that a new ice age might be on the cards. It had the longest run of snow covering the lowlands on record, a total of 67 days in England. It was also the coldest year of the 20th century.

As the winter wore on, the suffering grew as supplies of food and fuel ran low, and prices rose as goods became scarcer. Some 49 people died outdoors of the cold, many trapped in cars in blizzards. The legacy of that winter may not have been so profound as the devastation of 1947's winter, but it raised some interesting questions about what was going on with Britain's climate.

The freeze started to take grip towards the end of December 1962. Christmas Day was frosty and cold, with −9°C (16°F) recorded at Poole, Dorset. A policeman in Exeter had an unnerving experience when he heard a huge bang in a street. 'At that time villains still used explosives to attack safes, so I told the station what I had heard,' described Peter Hinchcliffe. 'But the cause of the bang had been the property on the corner of Southernhay East and Chichester Mews splitting from top to bottom, caused by a water leak that froze.'

It was the start of the Big Freeze. By Boxing Day thick snow swamped much of England and many people would not see their streets clear of snow until early March. Soon afterwards a blizzard tore through the south of England and Wales with snowfalls that changed the landscape out of all recognition, engulfing country hedgerows and burying whole buildings. In Dorset, two doubledecker buses were buried up to their rooftops. RAF and Navy helicopters flew dozens of missions dropping food supplies and lifting out the sick and injured. A farm on Dartmoor was cut off by mammoth snowdrifts and remained marooned until March.

The café on Deal Pier in Kent had to be shut after it was taken over by hungry seagulls that ate almost everything except for a malt loaf. At Herne Bay, frozen waters eventually stretched 2½ miles (4km) out to sea. Margate Pier was surrounded by ice and tugs battled against ice floes on the River Medway. By mid-February, ice extended along the north Kent coast for 7 miles (11km), gently rising and falling with the tide like something from the Arctic Ocean.

Pennine villages became completely isolated and sheep lost in the snow; 'the Peak District looks like the Alps' reported an RAC official in *The Times*. The railway line between Minehead and Taunton in Somerset was blocked by a train stuck in a snowdrift, another train in the area was abandoned by its crew, who took refuge in a farmhouse, and rail conditions in the West were so bad that priority was given to trains carrying food, coal, oil and petrol.

Salting roads proved ineffective because it was so cold, and if the ice and snow did melt, it re-froze later during sub-zero nights. Walking on pavements was particularly energy-sapping and hazardous, when partial thaws left menacing icicles hanging off buildings like spears, threatening to drop on pedestrians below. The largest icicle of all was probably 60ft (18m) long, hanging off Hardrow Force, in the Yorkshire Dales, the highest waterfall in England, when it became completely frozen.

On 6 January, dynamite was used to clear an avalanche blocking the railway from Edinburgh to Carlisle near Galashiels. The following day, the temperature at Grantown-on-Spey, Morayshire, sank to –22°C (–7.5°F). Even in the centre of London, a 15ft (4.6m) snowdrift was reported blocking Oxford Circus.

Power cuts began, made worse by a work-to-rule by the Electrical Trades Union. The scale of disruption to electricity and gas grew worse through January until eventually the national power supply collapsed. A build up of heavy hoar frost on insulators of National Grid equipment caused short-circuits and flash-overs and the power network broke down. The East Midlands was cut off from the North and South and there were widespread power failures.

On January 18, a blizzard virtually cut Scotland off from England. Abandoned vehicles were strewn across roads all over the country. Several trains were trapped in snowdrifts, and ferry services had to negotiate through waters teeming with ice floes. Seagulls were frozen in ice at Poole Harbour and Heathrow Airport (known then as London Airport) was closed when an airliner skidded off an icy runway. Hundreds of London buses were put out of action when their fuel froze. On 22 January a car drove across the frozen Thames at Oxford, and almost the entire length of the non-tidal Thames froze, as far as Teddington. But the river did not freeze in central London as in the olden days of frost fairs, because the waters were too fast and were warmed by effluents pouring out from factory and power stations.

Conditions in the uplands were ferocious. Two groups of climbers were killed by avalanches on moorlands in Lancashire. A dramatic and record-breaking rescue mission was launched on 20 January at Fylingdales early warning station on the North Yorks Moors, which had been completely snowbound for several days, trapping 283 civilian workers inside. A fleet of RAF helicopters rescued the entire staff in treacherous conditions. 'In one of the biggest airlifts ever to take place in Britain, workers were ferried in batches of seven or eight, in wind gusts up to 80 miles an hour,' described *The Times*. They were flown to Whitby, 12 miles (19km) away, but many found themselves marooned there by snowdrifts.

Frozen canals spelt the final death knell for freight carried on waterways. Already fighting for survival from competition on the new motorways, the winter of 1962–63 effectively killed off freight on canal boats. However, roads and railways were also struggling to remain open. On 5 February a Glasgow to Stranraer train became trapped in heavy drifts and the passengers had to be rescued by helicopter. Diesel engine trains suffered frozen cooling systems and boilers, and steam locomotives used as replacements, struggled with frozen water and coal supplies.

Serious water shortages struck various parts of the country as mains water supplies burst in roads, which were damaged up to 2ft (60cm) deep by frost. In Windsor, some 200,000 gallons of water were lost each day from broken mains, and Liverpool lost 8 million gallons of water each day. Flooding from burst mains in London led to firemen called out to 1,473 flood incidents over one weekend. Thousands of people relied on tankers or street standpipes for their daily water supplies, but the standpipes were in danger of freezing up, so many were kept working with bonfires lit in the streets. A highway surveyor in Devon described the desperate plight of trying to keep essential services going. 'My men have been sheeted in ice. They have icicles hanging from their ears and gathering on their eyebrows. Telegraph poles have been iced to double their normal size and electricity cables have measured 6 inches in diameter.' Fractured gas pipes were another, and lethal, problem. In Salford five people died of gas poisoning inside their home when a gas main split open just feet outside their house.

Food supplies ran low and as a result the price of fresh foods rose by 30 per cent. Millions of milk bottles disappeared, smashed open by frost. Schools were closed by snow, freezing cold and lack of heat. 'School milk was frozen in the bottles. We broke the glass and held the frozen milk in our hands like lollipops. The ink froze in the ink wells overnight!', described one man of his schooldays that winter. Rubbish piled up in homes and streets as collections broke down, and, most gruesome off all, there were serious problems buring the dead without using pneumatic drills to dig graves.

Ports all over the country completely froze and in East Anglia explosives were used to release trapped ships. Menacing pack ice was drifting in the Mersey off Liverpool, in the Solent and Humber. At Herne Bay, Kent, the ice reached 2.5 miles (4km) out to sea. The Navy kept Chatham dockyard open by using an icebreaker although other London docks remained closed. Bradwell nuclear power station in Essex suffered a near disaster on 26 January, when there was a power cut and the station's cooling system froze. An engineer recorded: '5am. Station shut down due to grid failure. Diesels didn't work. Essential supplies failure!!! Control room in darkness!' None of the alarms sounded. The atmosphere inside the compound, for 15 minutes, was 'absolute mayhem'. That incident was not officially confirmed until 1988.

A 45lb lamb was roasted on Oulton Broad, Norfolk in a scene reminiscent of the old frost fairs. On 28 January 1963 a slow thaw took hold and set off a new hazard. Trains were diverted at Caerphilly after 10 tons of ice dropped from a ventilating shaft, and at Torpantau, Brecon, 50 tons of snow and ice overhung a tunnel mouth like a guillotine. Derbyshire County Council used 400lb of gelignite to blow up a snow cornice hanging 200ft (60m) above the Snake Pass, which had been closed to traffic between Manchester and Sheffield for 11 days.

The football season ground to a halt and was so badly disrupted that the football pools company were facing a crisis without any games to bet on, so they set up the Pools Panel with five experts to estimate the results for all the unplayed games.

Meanwhile, football grounds tried desperate measures to beat the freeze, including a tar-burner at Chelsea and flame-throwers at Blackpool. Halifax opened its football ground on 2 March to the public for ice-skating, with hundreds of people turning up. Not until March 16th – nearly three months after the big freeze started – was a complete programme of football played again, and the season was eventually extended to the end of May. There were widespread calls for underground heating to keep pitches free of snow and frost, which seemed like a good investment at the time. The Scottish Football League was so concerned by the frozen conditions that they proposed suspending the football season from early December to early March in future years.

The prolonged cold was leading to some strange sights in the natural world. There were reports of starving foxes battling with badgers over scraps of food and in Sussex foxes were reported hunting domestic cats. Some 70 per cent of the oyster beds off the Essex coast were killed by the icy conditions and millions of dead ragworms were found floating in the sea off the Dorset coast. Even deep sea cod, plaice and whiting began to die of the cold over much of the North Sea. But shoals of sole migrated from Dutch and German coasts to find relatively warmer waters off the east coast of England, where they provided a bumper catch for the trawlers that could get to sea. Arctic birds such as the snowy owl, a resident of Scandinavia, were seen as far south as Essex, an incredibly rare sight. By the end of the winter, the populations of native birds had been severely hit; the wren population fell by around 80 per cent and one bird spotter found more than 50 wrens huddled for warmth in one nestbox. Some species even faced near extinction, with the native Dartford warbler population down to just 12 pairs. It took several years for the bird populations to recover.

As in 1947, the cold came from air drawn down from the Arctic by a block of high pressure stuck around Scandinavia or Iceland. Depressions tracked southward of the British Isles, dragging fronts with snow across England, Wales and the southernmost parts of Scotland.

Strangely, one industry that should have revelled in the Arctic winter was left hopelessly high and dry. The nascent ski industry of Scotland was bereft of snow. The western Highlands of Scotland were so dry that Fort William recorded no snow at all and much of the countryside turned into a tinder-box, with moorland fires breaking out on the Isle of Skye. Scotland lay so close to the dominant high pressure from Scandinavia to Greenland, that although it had a cold winter the skies were incredibly clear and free of snow. The rest of Britain also enjoyed more sun than usual but interrupted by further belts of snow from southerly tracking depressions. Insult was added to injury when the cold continental air unloaded not only its snow but its dirt as well; bulldozers slicing through snow-drifts in a 3 mile (5km)-wide belt inland from Torquay revealed a dirty layer of soot that came from coal smoke in Germany.

However, on 14 February the first signs of a thaw had appeared in the West Country, and parts of Devon were flooded as weeks of snow and ice melted, leaving cars stranded by up to 4ft (1.2m) of water. On 15 February, water rationing began in Aberystwyth and Carmarthen. But for most places the thaw was gradual enough to avoid the sort of floods that devastated Britain in March 1947.

Overall, that winter dealt a severe blow to the economy in the snow, ice and powercuts. Economic activity dropped by about 7 per cent and unemployment rose as 160,000 workers were laid off. Output dropped £300–400 million over the winter, and construction was hit particularly hard, with housing construction falling by more than 40 per cent.

Furious arguments erupted about why Britain was unable to cope with snow when the continent managed quite well every winter. But that was itself part of the answer, because Europe was used to blocked roads and used expensive snow-clearing equipment. The 1963 winter goaded Britain into increasing stocks of heavy snow ploughs at strategic road and rail junctions, and increasing the numbers of heaters on busy railway-line points.

There were human costs to the winter. In the first four months of 1963, London's Australia House received more applications for immigration than in all of the previous year, and a record 55,000 Britons later left for Australia.

There were political costs as well. The UK relied heavily on electrical storage heaters at that time, in the years before North Sea gas came on stream, and Sir Christopher Hinton, chairman of the Central Electricity Board was forced to explain the limitations of the National Grid to supply enough power. The Parliamentary Select Committee on Nationalised Industries was not impressed, and it concluded that planning for the winter was poor. The forecasts of power supplies were too low to meet demand, and led to calls for much better power capacity. As a result the electrical supply industry made a huge investment, far more than was needed, with extra capacity that was not needed for another 20 years, an investment that cost billions of pounds. However, that additional power supply proved crucial in keeping the National Grid working in the defeat of the miner's strike in 1984.

But perhaps the greatest legacy of that terrible winter was a widespread feeling at the time that the climate was taking a downturn, and even heading for a new ice age. Global warming was unheard of outside academic circles, and instead there was a belief that the planet was heading for a new ice age, based on subtle changes in the Earth's orbit around the Sun. There were also concerns that aerosols from industrial pollution were accelerating that cooling by blocking out sunlight. Atmospheric scientists and the US Central Intelligence Agency (CIA) came together in an attempt to determine the geopolitical consequences of a sudden onset of global cooling, and *Newsweek* in an article in 1975 warned about the cold climate:

'In England, farmers have seen their growing season decline by about two weeks since 1950, with a resultant overall loss in grain production estimated at up to 100,000 tons annually.'

'There are ominous signs that the earth's weather patterns have begun to change dramatically and that these changes may portend a drastic decline in food production – with serious political implications for just about every nation on Earth.'

Against that backdrop it was small wonder that it was such a struggle in later years to convince politicians and the public about global warming.

Glasgow Storm

January 1968

On Sunday morning, 15 January 1968, one of the worst storms in living memory battered northern England and Scotland. In the hills at Great Dun Fell in Westmorland one gust reached an astonishing 134mph (215km/h). The 120ft (36.5m) radio mast on the fell, used for VHF transmissions to aircraft, was wrecked, but a temporary broadcasting system was set up.

But Glasgow felt the full brunt of the storm as the winds were squeezed between the surrounding hills. 'Gales ripped through the city causing havoc reminiscent of the blitz. The city was being treated as a disaster area,' described *The Scotsman*. At Glasgow Airport the roof of a hangar was ripped off in winds gusting up to 103mph (165km/h), and on the Clyde a lifeboat house and its lifeboat were flipped upside down and a dredger sank and its crew of three lost. In the city, three churches were demolished, the front of four flats was blown off, killing four people in one tenement, and others had an incredibly close escape just before another tenement collapsed, crashing onto a smaller house below. One couple had a close escape when a timber beam fell just over their bed but the beam saved them when it took the weight of a huge chimney that crashed down on top of it. Firemen with hydraulic equipment eventually freed the couple. A 100ft (30m) high crane crumpled under the force of the wind and a 500-ton dockside crane collapsed at Greenock docks. Altogether nine people were killed in Glasgow, 1,700 were left homeless, and a staggering 100,000 homes were damaged.

Across Scotland, 19 people died, mostly from chimneys crashing through roofs of tenements. It was the worst gale ever recorded in the UK in terms of losses to the residential housing. The storm exposed the desperately poor state of repair of Glasgow's tenements and led to a new era of regeneration grants.

Braemar

January 1895 and 1982

Braemar is a picturesque village nestling in the Cairngorms, some 60 miles (96km) inland from Aberdeen. It is the highest and most mountainous parish in the UK, more than 1,000ft (300m) above sea level, and surrounded by ski slopes. And it is also the coldest settlement in Britain.

The village and surrounding mountains was a favourite holiday retreat for Queen Victoria, who liked it so much she had Balmoral Castle built a few miles away. And in 1855 Prince Albert donated a weather station to Braemar, which was an inspired choice, because Braemar has a very special climate.

The village lies in a natural bowl in the mountains, and on a clear, calm winter night cold air slides down the slopes like water. On 11 February 1895 Braemar broke the record for Britain's lowest ever temperature, –27.2°C (–17°F). Braemar also logged –20°C (–4°F) or lower on nine other days that February, highlighting its vulnerability to Arctic freezes.

The winter of 1894–95 was bitter across Britain. Ice skating became hugely popular on frozen lakes, rivers and canals and there was even ice-yachting on Lake Windermere. The lakes in London were lit with flares at night for ice skating.

The Thames froze from Richmond upriver, with an ox roasted on the ice at Kingston and four-horse coaches crossed the river at Oxford and Wallingford. Although the Thames could no longer freeze over in London after the opening of the new London Bridge, severe winters could still play havoc with the working of the Port of London, with huge ice floes floating in the water. The lighters and barges on which the port depended were trapped, and the wharves came to a standstill, with thousands laid off work.

In an age that depended on coal for heat and power, the big freeze brought the nation to its knees. Vital coal supplies could not get through on railways blocked by snowdrifts and canals frozen solid, including the Manchester Ship Canal only opened the year before.

Homes were left cold and industry was crippled. Farmwork was impossible on fields covered by up to 4ft (1.2m) of snow for weeks on end. Unemployment soared, and with no welfare many had no other resort but charity soup kitchens for their only food. At Chatham, Kent, a news report described the desperation: 'The scene today was extremely touching: the hungry poor assembling in great crowds and fighting their way to the doors in their struggles to get bread.' Many people died of hypothermia, and starvation set in even though soup kitchens were set up for the poor.

*

The strange thing is that Braemar equalled its own record cold temperature again on 10 January 1982. 'Braemar Colder Than The South Pole', exclaimed a headline in *The Scotsman* newspaper, pointing out that the temperature at the South Pole that day was only −21°C (−6°F).

For some days, buses had been unable reach the village in the snow and the main street was littered with stranded cars, even though local farmers used snowploughs attached to their tractors to try to keep roads open. 'There is a touring bus stuck in the village with frozen diesel fuel, and they have been lighting a fire under it to try to get it going,' reported John Stammers, of the Braemar Mountain Rescue Association in an interview with *The Scotsman*. 'We have been thinking that when God made whisky it was a good job he put plenty of alcohol in it.'

That night the temperature plunged to its record low in the clear, calm weather. 'It's one of those records which you would like to have and one which you would wished you did not,' he said. 'Half the village is without water and if it gets much colder there is no telling what damage the frost will do to the houses.'

The rest of Britain was gripped in the bitter freeze. Temperatures fell to −17°C (1.5°F) in Edinburgh and Aberdeen, and to −14°C (7°F) in Glasgow. On the Firth of Forth, fishermen returning to harbour reported fish freezing instantly as the nets were hauled in. Two people died of cold in England, one a farmer going out to feed his cattle in Shropshire, and another went missing.

Dozens of cars were abandoned along the M4 motorway, and mountain rescue teams worked to take people trapped in their vehicles to safety. At one point police rolled snowballs on to the motorway sliproads to stop motorists who ignored 'closed' signs at the entrances to the motorway.

The snow was the worst in memory in Wales. Hundreds of schools were closed for days and bread, milk and other basic food supplies ran out. Territorial and regular Army units mounted 'Operation Snowman' to help people stranded by huge snowdrifts. South Wales was hit particularly bad, and 500 steelworkers were marooned for five days at Port Talbot and Llanwern steel works, South Wales, although they worked round the clock to maintain fires to protect the furnaces from cooling down dangerously. In Cardiff, blizzards made every street almost impassable and one of the city's most famous buildings, the Sophia Pavilion, was destroyed when its roof collapsed under the weight of snow.

Strange Places to be Warm in Winter

Where is the warmest place in Britain in winter? Some sheltered cove on the South Coast, or perhaps an urban heat island like London? However, the highest temperature record in the depths of winter belongs to Abergwyngregyn, often shortened to Aber, on the North Wales coast, a historic village with far-reaching views over the Irish Sea and even the Isle of Man visible on a clear day. Towering above the village, to the east, stands a rocky outcrop Maes Y Gaer, believed to be the site of an Iron Age hill fort, and further south is Snowdonia.

On 27 January 1958, Aber reached 18.3°C (65°F), the highest January temperature recorded in the UK. And it repeated the exact same temperature again on 10 January 1971. Perhaps

North Wales should be a warm winter holiday destination instead of the Canary Islands.

The secret of this hot spot is a lesson in geography. The mountains to the west and south of Aber are generally bitterly cold in winter and one of the wettest regions of Britain. It would seem to be the most unlikeliest place to record any warmth. As depressions roll off the Atlantic, they crash into the mountains and pour with rain. But once the winds have passed over the mountains they sweep down the other side and warm up, an effect called the fohn wind. It works rather like a bicycle pump: as the air gets squeezed it warms up. It is a phenomenon better known in the Alps, where the warm, dry fohn winds send temperatures soaring, melt snow and turn wooden buildings into dangerous tinderboxes.

Aber is not unique, because all along the North Wales Coast often basks in the fohn heat. Hawarden on the eastern fringe of North Wales, close to the English border at Chester, is a historic village where William Gladstone and Michael Owen used to live. Hawarden also holds the distinction of the highest temperature ever recorded in Wales: 35.2°C (95.4°F), on 2 August 1990.

Freak warm mountain winds have hit the far north of Scotland as well. On 17 February 2003, the tiny village of Altnaharra in Sutherland recorded 11.5°C (52.6°F), more like a balmy spring day than winter, by far the hottest place in the UK and also warmer than European hotspots such as Barcelona, Athens, Corfu and Crete. Altnaharra also recorded air as dry as the Sahara Desert. It was caused by a fohn wind blowing over the Highlands, sweeping up from the south and forced over the mountain tops. Michael Mugford, owner of the Altnaharra Hotel, described the remarkable conditions. 'It's unbelievable, it's like you're in Greece the sun is so bright and warm, and we've got no snow whatsoever.' The balmy conditions were very confusing for the local wildlife. 'It's very strange – immense numbers of birds are singing, even the owls are unusually active' explained Michael Mugford. 'We've also had about 40 stags rutting in the garden, which is definitely not the right season for them because the hinds are in calf.'

Strange Things in the Stratosphere

1996 and 2008

Across the UK on 16 February 1996, fabulous streaks of psychedelic clouds appeared in the evening sky, shimmering with a strange metallic lustre, in the same way you often see colours on a film of oil floating on water. It was certainly an experience unlike anything else in the sky.

These were mother-of-pearl clouds, named after their fabulous iridescent colours. They float high up in the stratosphere, way above ordinary clouds, and are seen after sunset or before dawn when they blaze with unbelievably bright colours. They are filmy sheets slowly stretching and contracting in the twilight sky and are so bright because at some 12 miles (20km) high in the sky they are lit by the sun hidden below the horizon. These clouds are seen mostly during winter in the Arctic, and only rarely appear as far south as Britain.

Another display of mother-of-pearl clouds was seen across Scotland and northern England on 30 November 1999. This time it came in early dawn, as thin saucer-shaped clouds across the sky glowed with iridescent reds, pinks, oranges and turquoise. Then as the sun rose over the horizon the entire multicoloured display vanished.

Mother-of-pearl clouds are composed of tiny crystals of nitric acid. As sunlight glances off and around these crystals they shimmer with colours.

Another spectacular stratospheric phenomenon was seen over much of Britain a few evenings in mid-February 2008. Long after the sun had sunk below the horizon, the sky blazed with a magnificent vivid orange glow, the colour of a tangerine. This, too, was caused by stratospheric clouds made up of crystals of nitric acid, but without the form and all the colours of mother-of-pearl clouds.

Clouds are usually rarely seen in the stratosphere because it is intensely dry. But when the stratosphere is at intensely cold temperatures, below −70°C (−94°F), acid ice crystals can form. Such clouds usually form only in the extreme cold of winter in the Arctic or Antarctic, but in 2008 there was exceptionally cold air in the stratosphere over Europe, and so the stratospheric clouds appeared.

Despite their phenomenal beauty, these unusual clouds are a signal of something quite menacing in the stratosphere. The tiny acid crystals that shimmer with such stunning colours also drive chemical reactions in the stratosphere that eat into the Earth's ozone layer, the barrier that protects us from lethal barrages of ultraviolet rays from the Sun. Fortunately such conditions rarely last more than a week in this country, but in the Arctic and especially in the Antarctic they last for weeks and destroy so much ozone that massive 'ozone holes' appear that allow dangerous levels of ultraviolet light to reach the Earth's surface.

Diamond Dust

February 2008

Strange things were seen on the morning of 20 February 2008. It was a cold but stunningly clear and sunny day across much of the UK, when strange things were reported floating in the air. It looked as if showers of glitter had been thrown into the air, a gentle shower sparkling in the bright sunshine.

It is not often that things fall out of a clear blue sky, especially in calm conditions with hardly a breeze in the air. But it was the cold air that gave the secret the away, because this was a phenomenon more often seen in the Arctic than Britain. It is called 'diamond dust', a glamorous title that exactly conjures up the tiny sparkling crystals of ice that flutter in the air. Even though there was a clear blue sky, the air was cold enough that its moisture froze into tiny ice crystals.

In polar regions, people have been known to walk through clouds of the ice crystals, leaving behind a tunnel in the shape of their body. Usually diamond dust forms at exceptionally cold temperatures, around –30°C (–22°F) in clean air, but if there is dust or pollution in the air it can form at higher temperatures as moisture crystallises on the particles.

But in northern parts of Britain, that day began in fog and another peculiar sight was seen. Snow tumbled out of the fog and covered the ground in a fine, crunchy layer of whiteness. These were clumps of ice crystals, smaller than ordinary snowflakes. The strange thing is where the fog came from, because the air was incredibly dry. It seems likely that pollution pumped out of power station cooling towers and other industries became trapped in cold air close to the ground, held down by a lid of warmer air higher up. In fact, steamy plumes from central heating systems could be seen trapped just tens of metres off the ground. The fog crystallised in the cold, something known as 'industry snow'.

Other parts of Britain woke up to a landscape covered in what looked like icing sugar. This was rime frost, a white, feathery frost that forms on very cold, foggy nights. In this case, the tiny droplets of water in the fog instantly turned to ice as soon as they touched any freezing cold object, such as trees and fences, and built up into a thick layer of white frost.

It was, indeed, a remarkable morning of winter weather.

Rhubarb Growing Too Warm

The rhubarb growers of Yorkshire could be facing ruin because of climate change. The area between Leeds, Bradford and Wakefield is one of the finest rhubarb-growing areas in the world thanks to its cool climate and fine soil. Its most prized crop is a delicacy called 'forced rhubarb', an early season variety that needs a substantial cold spell in the autumn to break its dormancy. Only then can it be taken into dark, warm sheds to 'force' it to grow at a rate so rapid that you can actually hear a popping sound as the sheath covering the crown of the leaf opens up.

The darkness stops the plants photosynthesising so that the sugar goes to the stems from the roots, and they also grow twice as fast. The end result is delicately flavoured rhubarb without the stringy bits.

To maintain darkness, the forcing sheds are only lit by candlelight – any other light will stop the growth of the rhubarb and scupper the harvest. In fact, the dark forcing sheds have become a major tourist attraction, with a Rhubarb Festival in Wakefield at the end of January to the beginning of February. By this time the plant is ready to harvest.

In fact, rhubarb is very much a plant of the cold. It originally came from Siberia and when it was brought here in the 1600s it thrived in our damp cold climate, especially in the West Riding of Yorkshire. There the rhubarb industry grew so big that a local railway was built for the trade and a nightly train called the 'Pink Express' delivered the prized crop to London.

Traditionally, the rhubarb roots used to be ready for forcing around 1 November, but now the autumn weather is so mild that the process has been delayed as late as Christmas, losing the early advantage over competitors. The fear is that the climate is turning so warm that the Yorkshire rhubarb crop won't be able to break its dormancy at all, putting the entire trade out of business.